Praise for *The Infinite Staircase*

"Geoffrey Moore has crossed the chasm from marketing to metaphysics. The result is breathtaking."

—**Guy Kawasaki**

"For many years I've had the benefit of sage advice from Geoffrey Moore, one of the world's most influential business strategists. In this profound, remarkable work he explores the meaning of human existence and opens our minds to a new paradigm for understanding our place in the universe and a set of strategies for living an ethical life."

—**Marc Benioff, chair and CEO of Salesforce**

"At IDEO we apply design thinking to our client's toughest challenges. In *The Infinite Staircase*, Geoffrey Moore has used his own design thinking skills to take on the challenge of embracing a science-based understanding of how life emerged on Earth and finding within it the foundations for leading an ethical life. It is a *tour de force* in its own right and a book for our times."

—**Tim Brown, chair of IDEO and author of *Change by Design***

"Who are we? The answer to this question, the central question of the humanities, has prompted many different answers. None of the better-known answers to that question, as Geoffrey demonstrates, are wrong; each is merely incomplete, a piece of the puzzle. With this puzzle assembled, Geoffrey turns to the second critical question: How should we behave—and why? He then deftly points out that the answer to this question flows from the answer to the first: ethical behavior is advantageous to us as individuals and necessary to us as societies, and does not require the added directive of religion.This is a daring book, to be sure—a dare that Geoffrey has answered with his trademark concision and clarity. And like all daring acts, it produces a thrill—one for which I always will be grateful."

—**Harry Beckwith, *New York Times* bestselling author of *Selling the Invisible***

"Geoffrey Moore's *The Infinite Staircase* is an elegant exploration of grand questions that will change how you see the world. To the mix of physics, business, and history found in *Loonshots*, Moore has added biology, chemistry, linguistics, and ethics—a spectacular combination."

—**Safi Bahcall, international bestselling author of *Loonshots***

The
Infinite
Staircase

Also by Geoffrey A. Moore

Crossing the Chasm

Escape Velocity

Inside the Tornado

The Gorilla Game

Living on the Fault Line

Dealing with Darwin

Zone to Win

The
Infinite
Staircase

**What the Universe
Tells Us About Life, Ethics,
and Mortality**

Geoffrey A. Moore

BenBella Books, Inc.
Dallas, TX

BenBella

BenBella Books, Inc.
10440 N. Central Expressway
Suite 800
Dallas, TX 75231
benbellabooks.com
Send feedback to feedback@benbellabooks.com

BenBella is a federally registered trademark.

Printed in the United States of America
10 9 8 7 6 5 4 3 2 1

Library of Congress Control Number: 2020058097
ISBN 9781950665983 (trade cloth)
ISBN 9781953295378 (ebook)

Editing by Glenn Yeffeth
Copyediting by Karen Wise
Proofreading by Sarah Vostok and James Fraleigh
Indexing by WordCo Indexing Services
Text design by Publishers' Design and Production Services, Inc.
Text composition by Katie Hollister
Cover design by Sarah Avinger
Cover image © Shutterstock / Bulatnikov
Printed by Lake Book Manufacturing

Special discounts for bulk sales are available.
Please contact bulkorders@benbellabooks.com.

To Marie,
the love of my life,
and
to the family we both love so much:
Margaret, Daniel, Alayna, Michael,
Anna, Dave, and Noah

Contents

Preface

Let me say at the outset that this book is a departure from my prior published work. By profession, I am a business strategist serving the high-tech community. I have the privilege of assisting Microsoft, Salesforce, Intel, and other companies both large and small in figuring out how they can best compete and succeed in a rapidly changing marketplace. My first book on strategy, *Crossing the Chasm*, has sold over one million copies since its initial publication in 1991 and is still required reading in many business schools and entrepreneurship programs. Subsequent books have allowed me to continue my practice as author, speaker, and adviser to the present day. It has been a great ride, and I have no intention of stepping down anytime soon.

In that context, you might be surprised to learn that I started my career as a professor of English and a Renaissance scholar. My first book-length effort—foreshadowing the work you have in hand—was a PhD thesis titled *Strategies for Living*. It was an analysis of Edmund Spenser's *The Faerie Queene*, an epic poem first published in 1590. Spenser offered it as a gift to Queen Elizabeth I, explaining that his aim was to "fashion a gentleman or noble person in virtuous and gentle discipline." My intent for the book you have in hand is not all that different.

Like any good strategist, Spenser derived his advice about how people should behave from his understanding of how the world works. That understanding was anchored in a conceptual model called the Great Chain of Being, an idea that held sway in Western civilization from Classical antiquity to the Early Modern period. In it, God is positioned at the apex of creation, from whence descend all the different orders of beings, beginning with various classes of angels, descending further to mankind, then to animals, other living things, and finally down to the material elements of earth, air, fire, and water. The resulting worldview was one based on a hierarchy of value, with each link in the chain having its assigned place, all leading to a social, political, and economic order that was both stratified and conservative.

That order persisted into the Renaissance, but by the 17th century, it had become so encumbered by numerous unjust and oppressive entitlements that it spawned a series of rebellions. These began with the Enlightenment, followed later by Romanticism, and later still by Modernism. Each of these movements distanced itself further from a legacy of religious narratives, replacing them with increasingly secular story lines instead. Hierarchies were still acknowledged, but they were no longer divinely authorized. Instead, rather than having order imposed from above through intelligent design, it was increasingly seen as spontaneously emerging from below, evolving into layers of increasing complexity, driven by impersonal forces.

How such a thing could possibly happen is as miraculous a tale as has ever been told. Today the story is more or less complete, but it exists in pieces, captive to various academic disciplines, accessible primarily to specialists in those fields. The goal of this book is to pull those pieces together into a single coherent story using everyday language appropriate to a generalist, intellectually curious reader.

In this context, the first part of *The Infinite Staircase* is, in effect, a contemporary riff on the Great Chain of Being. It seeks to explain via the metaphor of the staircase how all reality is indeed structured as a hierarchy. Unlike the Great Chain of Being, however, both the top and bottom of this staircase are shrouded in mystery—hence the infinite staircase. Fortunately, however, the middle parts are clearly in view, and that is where our story takes place.

Telling this tale will take up the first two-thirds of this book. In so doing, it will set the stage for the remaining third. There we will address the question, If this is indeed what the world is actually like, what does that mean for how we should act? What, to bring things back to my perennial concern, should be our *strategy for living*?

In a purely secular universe, this is not an easy question to answer. Absent a religious narrative, where does morality come from? Where does justice come from? How are we supposed to navigate our daily lives, with what moral compass, by what North Star? Faith, spiritual support, and authorized ethics are foundational to success in life. We cannot do without them. Traditionally they have been provided by religion. Can they be accessible through any other means? How could you find faith *within* science instead of *apart* from it? How could you secure spiritual support independent of religious belief? And absent a God at the apex of creation, what can authorize us to say that any given act is good or evil?

These are all questions we must come to terms with if we are to embrace a secular narrative and still hold ourselves accountable to our traditional values. We need those values to conduct our lives in honorable, useful, and coherent ways. We cannot conduct the business of living without them. What the universe can tell us about how to secure them, and how to integrate them into our daily lives, is the subject of the investigation that follows.

Geoffrey Moore
August 2020

PART ONE

Metaphysics

CHAPTER 1

The Purpose of Metaphysics

The purpose of metaphysics is to give the most comprehensive and accurate answer possible to a very simple question: *What is going on?*

It is an interesting question. We are all born onto this planet to live for some time, but not forever. At various points in our journey we stop to look around, assess our circumstances, interpret our situation, and reflect on any number of big questions like *Who are we? How did we get here? How did all this other stuff get here? How are we supposed to make sense of all this? Who is in charge? Is there a plan? What does it all mean?*

Many people alive today, myself included, were born into a religious tradition that provides time-tested answers to just such questions. In our coming of age, however, many of us rebelled against those answers such that we entered into adult life as seemingly free agents. Of course, we are much more bound to our roots than that, something that others see even when we don't. In any case, one way or another we forge ahead, picking up bits of wisdom along the way, and after a while we do develop a sense of how the world works, of what we think is going on. We may not have fleshed out all the bits, but we are pretty sure about the broad outlines.

This would all be quite reassuring but for the extraordinary number of people who hold views on this very subject that are strikingly different from ours. Marxists, for example, believe that the means of production are foundational in that they dictate the norms of society. By contrast, Enlightenment traditionalists seek those very same foundations in reason, and Romantics in a transcendent spirit. Meanwhile, Cartesians advocate for two separate realities—matter and mind—with materialists asserting that it is only matter that matters, and idealists that it is only mind.

How is it possible that all these people have got things so mixed up? Or could it be us? Are we the ones who have got it all wrong? But if that is the case, then why don't all these other people at least agree with one another, which they clearly do not? Maybe we all have got it wrong. Maybe it is impossible to get it right.

This is the state of mind that Hamlet finds himself in halfway through Shakespeare's play, and that Rosencrantz and Guildenstern find themselves in halfway through Tom Stoppard's wonderful parody of the same. Like many in our contemporary world, they find themselves paralyzed by anxiety and doubt. But at this very moment there is only one salient point to register: as long as we linger in this state, we haven't answered the question! *What is going on?* Everybody agrees that *something* is going on, that it involves us, that it has consequences, and that it is happening here and now. Given that, sitting things out is not really a viable option, so we had best just suit up and get ourselves back into the game.

In that spirit, let me give you the game plan for the first part of this book. We are going to develop and explore a metaphysics, an explanation of what is going on, organized around a *hierarchy of systems*. Two thousand years ago this hierarchy was called the Great Chain of Being, something that was understood to reach from inanimate stones at its base, up through all the various forms of life, and then to humankind, and then on to a host of celestial beings, finally achieving its highest point in God. Two thousand years later, we will invoke the metaphor of the infinite staircase to describe something analogous. In the modern secular view, our staircase reaches from the smallest detectable particles in the quantum universe, again up through all living things including humankind, out to our solar system, our galaxy, and the billions of other galaxies, adding up to the totality of the universe itself.

As contained as it may seem, this staircase is infinite in several dimensions. First of all, it has no top or bottom, at least not as far as we can verify. That is, no matter how elementary the particles we are able to detect, we can never be sure we have reached absolute rock bottom. And no matter how far out into the universe our sensors can reach, we have never come close to thinking we have reached the edge. Moreover, even the stairs we have in plain sight stretch out in multiple directions, such that at every level there remain new dimensions to explore, new discoveries to be made.

Fortunately for our purposes, the parts of most immediate interest to us are huddled conveniently right here on Earth and open to direct inquiry. In particular, it is the living parts that attract our greatest interest. These consist of a nested set of systems that begins with the microbiology of individual cells and reaches up through tissues and organs to organisms, and further to populations and ecosystems, eventually reaching all the way up to Earth as a whole. The goal of our staircase approach will be to divide up our local landscape into recognizable levels of reality, to account for the dynamics of each level as accurately and completely as possible, and to organize the whole of reality as a hierarchy of levels.

Our metaphysics is organized around two principles. The first of these principles is *nesting.* A nested system is made up of component subsystems, which in turn are made up of their component subsystems, and so on, all the way down. Thus, for example, this book is made up of chapters, which in turn are made up of paragraphs, which in turn are made up of sentences, which are made up of words, which are made up of letters. At the same time, in a nested system, each component is also itself a subsystem of a higher-order, more complex system, which in turn is a component of some yet still higher-order, more complex system, and so on, all the way up. Thus, a person is a member of a family, which is a member of a community, which is a member of a state, which is a member of a nation, and so forth. As we have already mentioned, while the very top and the very bottom of our hierarchy of nested systems are shrouded in mystery, the middle stairs are clearly in view. By leveraging the discoveries of the past half century, we can develop a remarkably rich and reliable representation of all of reality that we can detect. This is the great metaphysical achievement of the twentieth century.

The second principle that frames our contemporary understanding of the universe is *emergence*. A system is emergent if it arises from the interaction among component subsystems in ways that are not directly determined by them. In other words, its emergent properties are neither reducible to, nor deducible from, the properties of its components. This is true of beehives, as well as traffic jams, and clouds for that matter. It is true of companies, of industries, and of nations, as well as being true of you and me. In every one of the cases, novel phenomena have simply emerged into being.

Frankly, this is nothing short of miraculous. Where can all this order be coming from? For most of human history, the obvious answer was from a divine creator, be that the personal God of the West or the impersonal field of Being of the East. Over the past several centuries, however, an alternative answer has been developed, anchored in the concept of *self-organization*. This concept allows us to account for the emergence of complex ordered systems without having to posit a creator. But how exactly does it work?

Self-organization is easiest to grasp in the context of agent-based components, be they ants, or cells, or robots. Each agent is pursuing its own agenda, one that manifests itself in the agent's local behavior. At the same time, however, each agent is also interacting with its peers. In situations of emergence, surprisingly, these micro-level interactions do not result in random outcomes or chaos, but rather in a macro-level phenomenon that has its own persistent reality. Every cell in your body, for example, has a local agenda. It does not "know about" any of your other cells, although it "hears from them" via chemical and electrical signals. From its point of view, however, these signals are simply part of the environment, external to its own operations. Nonetheless, these very same cells self-organize to make up your tissues, just as your tissues self-organize to make up your organs, just as your organs self-organize to make up your body. In every case, there is a higher level of organization—tissues, organs, body—that is made up of component elements: cells, tissues, organs. This higher-level entity represents a whole that is not just greater than, but also behaviorally distinct from, the sum of its parts.

Now, we call this self-organization because it unfolds without direction from a centralized control function. That is, there is no architect or contractor working off a master plan. Instead, there is a community of independent agents operating within a set of positive and negative feedback loops. As these feedback loops interact, a kind of equilibrium emerges in which the system as a whole stabilizes around a repeatable, but never identical, set of behaviors. You can see this in something as simple as geese flying in a V formation or as complex as how prices stabilize in free markets.

In chaos theory, such states are called "strange attractors." You can imagine them as depressions in a landscape of possible "phase states," sort of like dents in a mattress into which self-organizing systems can settle. Once settled, they can be further driven by positive and negative feedback loops into future behaviors that are increasingly familiar, but never deterministically predictable. This is how tornadoes form, schools of fish assemble, and stock markets rise and fall.

Rather than being disrupted by the innumerable random changes they incorporate, self-organizing systems are actually able to take advantage of them to explore alternative phase states, increasing the likelihood that they will find ever-stronger attractors that will lead to ever-more sustainable persistent behavior. This realization has given new life to the materialist account of reality. Earlier versions of materialism were shaped by two traditions anchored in purely mechanical metaphors. The first of these is *reductionism*. It claims that all complex things can be explained in terms of the operations of their component parts. The second is *determinism*. It claims that every discernible effect can be traced to a specific identifiable cause. While both have made valuable contributions to Western thought, neither is able to account for emergence, nor explain how or why reality is hierarchically organized into levels, or where such levels might come from. That is the work we are undertaking here.

The Infinite Staircase

In the chapters that follow, we are going to take a linear path up through Earth's systems hierarchy from bottom to top, leveraging the principles of emergence and self-organization as best we can perceive them. At each step in our staircase, a higher level of complexity will emerge and take its place as the next defining level of organization in the hierarchy. Each of these levels is made possible by those below it, and, in turn, each makes possible those above. I am not claiming that this framework addresses every possible level, only that the ones it does address are essential, and that they are presented in their proper order. There is, in other words, one and only one sequence for the staircase, and it is as follows:

The Infinite Staircase
(as far as we can see)

11. Theory
10. Analytics
9. Narrative
8. Language
7. Culture
6. Values
5. Consciousness
4. Desire
3. Biology
2. Chemistry
1. Physics

The fundamental claim here is that each lower level in the staircase is prerequisite for the emergence of the stair above it. That means each higher level *entails* all the levels below it—they are necessarily included in its reality. In half the cases, this claim is pretty easy to grant:

- Chemistry necessarily entails physics.
- Biology necessarily entails chemistry.
- Desire necessarily entails biology.

- Narrative necessarily entails language.
- Theory necessarily entails analytics.

Hard to find much that is controversial there. But the same cannot be said about the other half:

- Consciousness necessarily entails desire.
- Values necessarily entail consciousness.
- Culture necessarily entails values.
- Language necessarily entails culture.
- Analytics necessarily entail narrative.

One can readily imagine a whole host of objections to any one of these assertions. Moreover, we can find further grounds for disagreement because the staircase is bidirectional, meaning it can be walked down as well as up. In that context, any item on a lower stair has no dependence on any item above it, resulting in some additional controversial claims:

- Culture does not entail language.
- Values do not entail culture.
- Consciousness does not entail values.
- Desire does not entail consciousness.

The reasons for spelling all this out are twofold. First, these are the claims that will organize each of the following three chapters, which, taken together, set forth our contemporary take on metaphysics.

- The first chapter will address the bottom stairs of physics, chemistry, and biology—the realm of materialism. Here we will develop a *metaphysics of entropy*, with the Second Law of Thermodynamics providing the organizing principle.
- The second chapter will address the middle portion of the staircase, the stairs of desire, consciousness, values, and culture—the domain of organic evolution. Here we will develop a *metaphysics of Darwinism*, anchored in the processes of natural and sexual selection.

- The third chapter will address the highest portion of the staircase, the stairs of language, narrative, analytics, and theory—the domain of ideas. Here we will develop a *metaphysics of memes*, anchored in processes of social selection that directly parallel natural and sexual selection.

By the end of that third chapter, if all goes well, we should have as comprehensive and accurate an understanding of how the world works as contemporary culture can provide.

The second reason for spelling out all these claims so starkly is to bring the whole domain of metaphysics into the light of everyday language and experience. *Metaphysics*, let's face it, is an intimidating word. *Physics* alone can be daunting enough. Add a *meta-* to it, and you can feel like you are biting off more than you can chew. And indeed, if you read very far into the history of the subject, you run into a whole cast of characters whose names you recognize but whose books you have never read. Consider a starting lineup made up of Descartes, Spinoza, Leibniz, Hume, Kant, Hegel, Russell, Wittgenstein, Kierkegaard, Heidegger, and Nietzsche. Need I say more? Each name radiates impenetrability, and collectively they form a defense as formidable as that of the 1985 Chicago Bears.

In my view, the impenetrability of metaphysics in the Western intellectual tradition can be traced directly to these authors and their peers violating one or more of the hierarchical rules that underpin our staircase. Typically, this happens when one stair in the staircase is privileged over all the others. Idealists over-privilege the stairs at the top, envisioning a world in which immaterial form imposes intelligible order on matter through the act of creation. Materialists over-privilege the stairs at the bottom. For them, matter is the ultimate reality, and everything else is an inconsequential side effect, what philosophers call an *epiphenomenon*. Both groups are at odds with the core thesis of this book: the fundamental basis of reality does not reside on any one stair but rather in the staircase itself. It is the phenomenon of levels that matters most. We need to honor each level in its own right while at the same time seeing it in relationship to all the others.

Whenever this principle gets ignored, philosophers end up at cross-currents with reality as it actually is. In trying to reconcile their ideas with the contradictions that emerge, they turn to increasingly distorted acts of language that ultimately end up collapsing under their own weight. Look at all the countless words spilled over Plato's theory of Forms, or Descartes's mind-body dualism, or Kant's notion of *a priori* knowledge, with no resolution in sight. This is what gives metaphysics a bad name. Metaphysical clarity comes from putting issues in the right context—locating them on the right stair in the staircase—and philosophical confusion comes from failing to do so.

So, if nothing else, let this be a warning to us, or rather a warning to me and a litmus test for you. This metaphysical journey we are about to undertake should be easy. It should be simple to follow and reach conclusions that feel natural, even obvious. These will be signals we are on the right track. If the going should ever get tough, the tough should definitely get going—back to the last place that felt "right" in order to find some other, easier route forward.

And with that in mind we can turn our attention to the bottom of our staircase and what we are calling the *metaphysics of entropy*.

CHAPTER **2**

The Metaphysics of Entropy

We begin our journey up the infinite staircase in the domain of the material sciences. Here physics represents the foundational stair, with chemistry emerging from it, and biology emerging from chemistry.

The Infinite Staircase
(The Metaphysics of Entropy)

11. Theory
10. Analytics
9. Narrative
8. Language
7. Culture
6. Values
5. Consciousness
4. Desire
3. Biology
2. Chemistry
1. Physics

What unites all three is that they are all ultimately organized around entropy. Understanding entropy turns out to be the key to unlocking the mystery of emergence. That's why we are starting here. And because of the extraordinary advances made in this domain over the past century, we get to start with a bang!

What Is Entropy?

Every culture has its creation myth, and modern Western secular culture is no exception. Ours is called the Big Bang. Here's the story.

Some 13.8 billion years ago, as best we can tell, there was no time, no space, and no matter—just a lot of nothing, except that there was not even nothing. Nada. And then, *presto*, space and matter appeared simultaneously everywhere in the universe, all at once! That is, in a millionth of a millionth of a millionth of a millionth of a millionth of a second, the universe of space and matter expanded by a factor of a billion billion billion—give or take! From that jump start (they call it *cosmic inflation*), it has been expanding ever since. This universe, as best we can tell, consists of billions and billions of galaxies, each of which, like our very own Milky Way, contains billions and billions of stars, one of which happens to be our sun, around which orbits our Earth, a planet hosting a life form that, during its exceedingly brief existence, has somehow been able to figure all this out.

Now, I ask you, have you ever heard of any creation myth from any culture that is more astoundingly miraculous than that? Of course, being a secular culture, we don't call it a miracle—we call it a *singularity*—but, I mean, really.

Nonetheless, here is what is amazing about contemporary Western cosmology. Once you grant it this singularity, it is able to build an extraordinarily coherent narrative reaching from then to now, by leveraging the power of mathematics. This work was begun by the likes of Johannes Kepler and Galileo Galilei, elevated by Isaac Newton, James Maxwell, and Michael Faraday, and extended in wholly new ways by Albert Einstein and a host of others. Scientists can *calculate* this narrative and *verify* it experimentally. No other creation myth has ever come close to this performance. In its own way it is as miraculous as the Big Bang itself.

However, that still leaves us with a problem. You and I don't speak mathematics. We can't read it, we can't write it, and until it is translated into natural language, we cannot understand it. Moreover, a whole lot of what mathematics has to say is not germane to the topic at hand—metaphysics as it pertains to leading an ethical life. Some of it is, however, and the goal of this chapter is to tell that part of the story.

The hero of our story is a character named *entropy*. Entropy is typically understood as the irreversible tendency of ordered systems to become disordered. This is indeed a feature of entropy, but it is not the most important one, and overemphasizing it can result in a fundamental misunderstanding of how the universe works. The feature to focus on instead is that entropy is the irreversible tendency of energy to dissipate—for hot objects to cool, for energized molecules to find equilibrium in their lowest energy state. The universe, like an aspiring teenager, just wants to be cool.

Our understanding of entropy in relation to energy dissipation began in the 19th century with the study of steam engines and how they convert heat into work. The focus then was on the efficient conversion of energy, and the seminal insight was that 100 percent efficiency was impossible. There is no such thing as a perpetual motion machine. Some energy—indeed, in many cases, much of it—is inevitably dissipated and is thus unavailable to do work. This "lost" energy was called *entropy*.

Later in the century, scientists generalized these findings. They defined any system that converts heat energy into work as *thermodynamic*, and they described several laws that govern all thermodynamic systems, the first two of which concern us here:

- The First Law of Thermodynamics says that energy can be neither created nor destroyed, only converted from one form to another. Thus, whatever got released at the Big Bang is all we are going to get.
- The Second Law of Thermodynamics says that heat energy will flow spontaneously from a center of higher concentration (a hot body) to one of lower concentration (a cold body) and never the other way

around. Thus, the natural tendency of heat is to dissipate without doing any work other than to bring the environment into thermal equilibrium. Ultimately, taking this law to its logical conclusion, the entire universe will be at one uniform temperature. With no differences in temperature, no work can be done, and no life can exist. In essence, the universe will have finally wound down.

With this background in mind, note that *entropy* can have two distinct meanings:

1. *Energy that gets lost to heat.* Here entropy is describing an effect, a *result* of the Second Law.
2. *The irreversible tendency of the universe to wind down.* Here entropy describes a *cause*, and thus becomes a shorthand way of referring to the Second Law itself.

Both senses are used in the material that follows, but it is primarily the second definition of entropy, the irreversible tendency of the universe to wind down, that is core to our argument. At that core is a paradox. Entropy, the tendency of everything to wind down, turns out to be the very engine that winds everything up! This is counterintuitive to say the least, so stay with me through this next bit.

The key point is simple. While the universe is indeed proceeding inexorably to dissipate all of the Big Bang's energy eventually, some amount of that energy is being diverted to do work in the meantime. And that meantime is a very long time indeed—13.8 billion years and counting. Moreover, the amount of energy being diverted is unimaginably large, enough to order all matter everywhere and to drive all motion from now until the end of time, including the matter and motion that constitute life on Earth.

All living things are themselves thermodynamic systems. We all ingest food, the energy of which can be measured as calories or units of heat, which our metabolisms convert into the work of building and maintaining our cells and intercellular structures. This work consists entirely of chemical reactions, every one of which is governed by the Second Law.

The chemical reactions that support life fall into two camps. Some of them are "favorable," meaning they occur spontaneously. Favorable reactions "run downhill," meaning they need no further energy input to get going. They just go with the flow to achieve a state of greater entropy. The products of these reactions are at a lower overall level of energy than the original reactants, with the difference in energy being released as heat. When iron rusts or wine sours, when acid burns your skin or food rots, these are favorable reactions driven by reactants that need no further encouragement. No work is being done, and no order is being created. This is indeed the universe winding down.

In "unfavorable" reactions, by contrast, the reactants are not predisposed to interact. For such reactions to proceed, they have to "swim upstream" first; work has to be done. This happens, for example, when nucleic acids are strung together to make DNA molecules or when amino acids are strung together to make proteins. Some external source of energy must be tapped to excite the reactants into a more energetic state, so they are disposed to interact, thereby leading to a completed reaction. The result is that one or more of the products of an unfavorable reaction will typically embody *greater* order than there was at the outset. In such cases, local to the domain of the reaction, there is actually a *reduction* in entropy.

Unfavorable reactions that decrease entropy locally must occur or life cannot exist. However, they can be brought about only by adding energy from the outside. Where does that outside energy come from? Well, ultimately the overwhelming bulk of all energy on Earth comes from the sun. After all, it too is doing its best to cool down, and shedding its energy is its only path. Here on Earth, plants convert that energy into the sugars, proteins, fats, and other biomaterials needed to build and maintain themselves. Some animals get their biomaterials from eating those plants, and others from eating those animals. In all living organisms, the energy used to maintain life comes from converting food into fuel.

And this is where the paradox of entropy gets explained. Decomposing food into fuel adds more entropy to the universe than is subtracted by the work done to create order. How do we know this? Because it is impossible for any thermodynamic system to be 100 percent efficient. Some of its energy *has* to get lost to heat. Thus, when you add up the bill globally, combining the local reaction where there was actually a reduction of entropy with the fueling reaction where there was an increase in entropy, the increase always exceeds the decrease. There will always be more total entropy at the end of any reaction than there was at the beginning. This balance-of-payments principle is built into the Second Law, and it cannot be violated.

As already mentioned, this paradox is counterintuitive. Because a reduction in entropy is necessary to create order and life, entropy itself looks to be an opposing force, not a supporting one. It is natural, therefore, to associate entropy with death, as in the following quote that is widely cited online:

> The flow of energy maintains order and life. Entropy wins when organisms cease to take in energy and die.

This is simply wrong. The flow of energy that maintains order and life is *enabled* by entropy. To illustrate this point, imagine a small factory on the edge of a river—say, a mill for grinding wheat into grain. The water in the river is flowing from A to B, but the water wheel is able to divert some of that energy via a system of gears and shafts to do work—in this case, turning a grindstone. The mill's mechanical system is diverting a portion of the river's flow to use for its own purposes.

All living things do the same thing. The dissipation of energy that enables life is like a river flowing irreversibly to the sea. Life's metabolic systems are like mill engines tapping into this flow to create and maintain the structures that make up all living cells. The structures themselves are highly ordered, but they are created by diverting energy from its unceasing flow toward greater disorder.

Peter Atkins makes this point particularly succinctly in his *Four Laws That Drive the Universe*:

> Whatever structure is to be conjured from disorder, it must be driven by the generation of greater disorder elsewhere, so that there is a net increase in disorder in the universe.

This is the essence of the Second Law. Far from being the enemy of order and complexity in life, entropy provides the very mechanism by which it comes into being. Without that river, there is no mill. Without the ever-present contribution of entropy, there is no order, no life, no staircase.

Everything that follows in this chapter is focused on applying this one idea across the domains of physics, chemistry, and biology. They represent the bottom three stairs in our staircase model, the realm of materialism. Looking at the material world through the lens of entropy lets us account for all the order in the universe prior to the emergence of consciousness. As our staircase makes clear, materialism cannot represent the totality of metaphysics, but it does get us off to a very good start.

Stair 1: Physics

Entropy and the Aftermath
of the Big Bang

"As the universe expanded, the universe cooled, and matter formed."
This NASA-based entry on Space.com makes it all seem so straightforward and reasonable. Don't kid yourself. Before one second had
passed, all the hydrogen nuclei there will ever be were formed. It took
another 20 minutes for helium nuclei to show up, but then we got all of
them, too. But then it takes almost 400,000 years before those hydrogen and helium nuclei can leverage the electromagnetic force to capture electrons to form stable atoms, and a whopping 100 million years
before gravity can pull together those atoms into clouds and then into
masses so the first stars can begin to shine. Inside their fiery cauldrons,
immense pressure and intense heat will drive a process called nuclear
fusion. That is what will create the huge swath of remaining elements in
the periodic table. Later on, these get distributed across the galaxy when
stars explode as supernovas, and even heavier elements get created when
black holes collide.

The key point is that every one of these events—the emergence of
the first nuclei and then atoms, the emergence of stars, the eventual
distribution of all the elements—represents an enormous expansion of
space, and every one of those expansions is tied to "cooling down." *Cooling* is a thermodynamic term, and it happens when entropy has dissipated enough heat that any given region has a lower concentration of
energy than it had before. But so what? Why does cooling make such a
difference?

The answer is that there are four fundamental binding forces that
hold together our material universe: gravity, electromagnetism, and the
strong and weak nuclear forces. Each of these forces has finite strength.
If the temperature is high enough to agitate material beyond the threshold that can be overcome by that force, the force cannot take effect. It's

like trying to get a class of second graders to sit still the day after Halloween. But once things are allowed to cool down below that threshold, then the force in question can take effect, and order can emerge. The forces that hold water together, for example, take effect once it has dipped below its boiling point of 100 degrees centigrade, just as the forces that hold ice together do so once water has dipped below its freezing point of zero degrees centigrade. Once anything cools enough for binding forces to get to work, then get to work they do. But that still begs the question, Why?

Binding forces increase the order in the material they operate on by shedding unstable energy and coalescing around a stable state. When they do so, they lower the overall energy of the material itself—that is, they cool it even further because they take it to a more stable state that requires less energy to maintain. That frees the excess energy to go elsewhere, either to fuel work or to dissipate heat. Binding increases the entropy in the universe overall even as it decreases the entropy in the material it is operating on. This is the Second Law at work. The universe does not have a bias to create order, quite the opposite. But one of nature's ironies is that one way to increase disorder globally is to increase order locally.

Now, based solely on this model of energy diverted to do work, we can account reasonably well for matter and energy as they were understood at the end of the 19th century. In the twentieth century, however, matters took an amazing turn, going down to a stairstep below classical physics to explore a domain known as quantum mechanics. More than one hundred years later, even the best of physicists are not completely clear on what exactly is going on at this level in the systems hierarchy. Part of the problem is that natural language does not map to the data in readily understandable ways. We say, for example, that a wave and a particle represent two separate and distinct physical forms, and yet we say that light—indeed, any form of electromagnetic radiation—is both a wave and a particle. Or we say that things, by definition, have location and momentum, but at the quantum level we say you cannot know both the location and the momentum at the same time. Or we say that nothing in the universe can travel faster than the speed of light, but at the quantum level, particles can be entangled such that, even when separated by an arbitrarily large distance, a change in one simultaneously spawns a corresponding change in the other.

All these examples are impossible to explain using the normal resources of natural language. There is obviously something going on, and mathematics has made some headway in clarifying what that is, but those of us who do not speak that language are barred from participating. So, for the purposes of this book, classical physics represents the bottom stair of our staircase. It is not actually the bottom, but it does represent a practical limit. Our understanding of life, ethics, and mortality must be grounded in natural language. We can take it only to the limits that such language allows.

With that thought in mind, we will turn our backs on the quantum-physical stair below classical physics and look upward to the next stair: chemistry.

Stair 2: Chemistry

The Emergence of Emergence

Chemistry—specifically, the chemical reactions that bind atoms into molecules—represents the first observable appearance of emergence in our journey up the staircase. As we have already noted, emergence arises from underlying subsystems self-organizing to create entities that have novel properties not present in their component ingredients. A familiar example is the English language. If all you knew were the 26 letters of the alphabet, you could not account for the novelty of words. If all you had was a dictionary of words, you could not account for the novelty of sentences. If all you had was an English grammar manual, you could not account for the novelty of English prose. If all you had was a manual of English prose, you could not account for the novelty of *Moby Dick*. Once you gain an understanding of any one of these higher levels of organization, you can see how the underlying elements contribute to its creation, but only in hindsight. You have to first experience the novelty and do the retrospective analysis before you can grasp what is actually going on.

In the case of chemistry, given everything that we know or could know about sodium and chlorine, we still could not anticipate salt. It has emergent properties that are not present in either of its two elements. Similarly, given the apparent similarity between salt and sand, we could not anticipate that one would dissolve in water and the other not. We can explain such things after the fact, but only by adding new conceptual models to our account of the universe. Such additions occur at every step in the staircase. Taken all together, they represent the sum total of all our knowledge. That is what makes the staircase so fundamental to our inquiry into metaphysics.

Specifically, the transition from physics to chemistry begins with the emergence of the atomic elements from the component elements of protons, neutrons, and electrons. Atomic elements are the building blocks of all detectable matter. There are over one hundred of them, and they are beautifully arranged in a framework called the periodic table.

Periodic Table of the Elements

1	2	3	4	5	6	7	8	9	10	11	12	13	14	15	16	17	18
H 1 hydrogen 1.0079																	**He** 2 helium 4.0026
Li 3 lithium 6.941	**Be** 4 beryllium 9.0122											**B** 5 boron 10.81	**C** 6 carbon 12.011	**N** 7 nitrogen 14.007	**O** 8 oxygen 15.999	**F** 9 fluorine 18.998	**Ne** 10 neon 20.180
Na 11 sodium 22.990	**Mg** 12 magnesium 24.305											**Al** 13 aluminium 26.982	**Si** 14 silicon 28.086	**P** 15 phosphorus 30.974	**S** 16 sulfur 32.065	**Cl** 17 chlorine 35.453	**Ar** 18 argon 39.948
K 19 potassium 39.098	**Ca** 20 calcium 40.078	**Sc** 21 scandium 44.956	**Ti** 22 titanium 47.867	**V** 23 vanadium 50.942	**Cr** 24 chromium 51.996	**Mn** 25 manganese 54.938	**Fe** 26 iron 55.845	**Co** 27 cobalt 58.933	**Ni** 28 nickel 58.693	**Cu** 29 copper 63.546	**Zn** 30 zinc 65.39	**Ga** 31 gallium 69.723	**Ge** 32 germanium 72.61	**As** 33 arsenic 74.922	**Se** 34 selenium 78.96	**Br** 35 bromine 79.904	**Kr** 36 krypton 83.80
Rb 37 rubidium 85.468	**Sr** 38 strontium 87.62	**Y** 39 yttrium 88.906	**Zr** 40 zirconium 91.224	**Nb** 41 niobium 92.906	**Mo** 42 molybdenum 95.94	**Tc** 43 technetium [98]	**Ru** 44 ruthenium 101.07	**Rh** 45 rhodium 102.91	**Pd** 46 palladium 106.42	**Ag** 47 silver 107.87	**Cd** 48 cadmium 112.41	**In** 49 indium 114.82	**Sn** 50 tin 118.71	**Sb** 51 antimony 121.76	**Te** 52 tellurium 127.60	**I** 53 iodine 126.90	**Xe** 54 xenon 131.29
Cs 55 caesium 132.91	**Ba** 56 barium 137.33	57-70 *	**Lu** 71 lutetium 174.97	**Hf** 72 hafnium 178.49	**Ta** 73 tantalum 180.95	**W** 74 tungsten 183.84	**Re** 75 rhenium 186.21	**Os** 76 osmium 190.23	**Ir** 77 iridium 192.22	**Pt** 78 platinum 195.08	**Au** 79 gold 196.97	**Hg** 80 mercury 200.59	**Tl** 81 thallium 204.38	**Pb** 82 lead 207.2	**Bi** 83 bismuth 208.98	**Po** 84 polonium [209]	**At** 85 astatine [210]	**Rn** 86 radon [222]
Fr 87 francium [223]	**Ra** 88 radium [226]	89-102 **	**Lr** 103 lawrencium [262]	**Rf** 104 rutherfordium [261]	**Db** 105 dubnium [262]	**Sg** 106 seaborgium [266]	**Bh** 107 bohrium [264]	**Hs** 108 hassium [269]	**Mt** 109 meitnerium [268]	**Uun** 110 ununnilium [271]	**Uuu** 111 unununium [272]	**Uub** 112 ununbium [277]	**Nh** 113 nihonium [284]	**Fl** 114 flerovium [289]	**Mc** 115 moscovium [288]	**Lv** 116 livermorium [293]	**Ts** 117 tennessine [294]	**Og** 118 oganesson [294]

*Lanthanide series

La 57 lanthanum 138.91	**Ce** 58 cerium 140.12	**Pr** 59 praseodymium 140.91	**Nd** 60 neodymium 144.24	**Pm** 61 promethium [145]	**Sm** 62 samarium 150.36	**Eu** 63 europium 151.96	**Gd** 64 gadolinium 157.25	**Tb** 65 terbium 158.93	**Dy** 66 dysprosium 162.50	**Ho** 67 holmium 164.93	**Er** 68 erbium 167.26	**Tm** 69 thulium 168.93	**Yb** 70 ytterbium 173.04

**Actinide series

Ac 89 actinium [227]	**Th** 90 thorium 232.04	**Pa** 91 protactinium 231.04	**U** 92 uranium 238.03	**Np** 93 neptunium [237]	**Pu** 94 plutonium [244]	**Am** 95 americium [243]	**Cm** 96 curium [247]	**Bk** 97 berkelium [247]	**Cf** 98 californium [251]	**Es** 99 einsteinium [252]	**Fm** 100 fermium [257]	**Md** 101 mendelevium [258]	**No** 102 nobelium [259]

This table itself is a terrific example of emergence. It represents an underlying order that one could not have predicted. Indeed, it was not even discovered until the 19th century, and it was not until the development of quantum theory in the 20th century that it was properly explained.

The table is organized around two principles. The first is the size of the atom itself, which is a function of the number of protons incorporated into its nucleus. Atoms are presented in rows ordered by atomic number from left to right, top to bottom, beginning with hydrogen with one proton at the upper left and ending with oganesson (a human-made element) with 118 protons at the lower right. (I'll explain the two additional rows at the bottom of the chart in a moment.)

The second principle organizing the periodic table, the one that was absolutely mind-blowing at the time of its discovery, is that, although it is laid out in rows, it is actually the columns that generate its key insights. The columns "emerge" from the rows. Atoms in the same column react in similar ways with atoms in other columns. As we will discuss in a moment, it has to do with the arrangement of the electrons that complement each of these nuclei. But no one in the 19th century knew there was such a thing as an electron. They just were tracking down reaction affinities, and this ordering emerged.

What we know now, however, is that each of these atoms in its elemental state has the same number of electrons as it does protons. These electrons arrange themselves in concentric shells, each shell having space for a fixed number of electrons. There are seven shells in total, concentrically arranged. The number of electrons in each shell, going from the top row to the bottom, are 2, 8, 8, 18, 18, 32, 32. This corresponds exactly to the number of cells in each row of the periodic table. If you add up all the numbers just cited, you get to 118, the total number of atoms represented in the table. Note that the bottom two rows need to be 32 cells long, but that would make the table hard to print. That's why there are two "extra rows" at the bottom—think of them simply as "inserts." Finally, if you count the total number of rows, seven in all, that represents the maximum number of shells that atoms can support—there is no eighth. That is an amazing amount of information to come from one table.

Now, as I said, you have to hand it to the 19th century for coming up with this diagram well before anybody knew there was such a thing as an electron. So how did they figure it out? The key was observing that any given atom likes to react with some of the other atoms in the table but not with others. The periodic table, in effect, represents the universe of marriageable atoms organized by their predisposition to hook up with one another.

Atoms that have their outermost shell of electrons completely filled are located on the far right-hand column of the table. These atoms are extremely stable, meaning that they have no disposition to interact with other atoms. Helium is one example, followed by neon and argon. They are said to be *inert*, possibly because *celibate* was already taken.

Every other atom in the table, by contrast, has an outermost shell with one or more empty spaces. The atoms on the left side of the diagram are just starting a new shell. Think of them as having one or more "extra" electrons. Those on the right, by contrast, have almost completed a shell, but not quite. Think of them as having one or more "empty spaces" that want filling. This is why atoms on the left are predisposed to combine with atoms on the right. Both are in search of some configuration that would bring their outermost shells to a state of equilibrium. These are the electromagnetic forces that drive all chemical reactions.

OK, so what? What does this have to do with the metaphysics of entropy? It goes back to the universal tendency for everything in the universe to seek to cool down. For a molecule, that means getting to a state of equilibrium. In that state, the molecule has released all the free energy it can. It has no further disposition to do anything (something like a teenager, eyes closed, earphones on, lying on your living room couch). Now, if you input energy to excite the molecule (say, by asking your teenager if he or she has done their homework), then you will take it out of equilibrium, and it will engage with the world, but only so long as it takes to get back to equilibrium. Matter—all matter, everywhere in the universe—really just wants to chill.

Order in inorganic matter results from atomic elements interacting electromagnetically until they stabilize into their state of lowest free energy. Crystals are a beautiful example. Where does their exquisite structure come from? It comes from their atoms snuggling up to each other to get as comfortable as they possibly can, and then not moving thereafter. Stabilized matter can, in turn, be incorporated into higher-order systems, in part through subsequent chemical reactions, in part through the pressure of gravitational forces acting on it and its surroundings. Under the impact of these forces, geological artifacts like gorgeous mountains and beautiful lakes simply emerge. The universe doesn't care. All it insists upon is that these emerging entities pay their bills. That is, at the end of the day, there must always be a net gain in entropy. The river of entropy has to flow to the sea, regardless of how many mill wheels it may spin up along the way, even when those wheels spin up more wheels, and so on, and so on.

To sum up, then, entropy is the universal tendency to cool—to dissipate as much free energy as one possibly can, whenever one can. Thus, when you put two chemicals together, and they see a chance to get to shed free energy to get to a greater state of equilibrium, they take it. In some cases, you will have to prime the pump by adding some extra activation energy. This added energy takes the chemicals out of a local equilibrium in order to let them interact to create a greater, more global equilibrium. Once activated, they grab the opportunity to react provided they can end up in a more stable, less energetic state. The lower the free energy in any system, the less predisposed it is to react, and the more stable its current state. This is the river of entropy in action.

But wait: *What about life?* Organic matter is not like inorganic matter. It is continually building and repairing its internal structure for as long as it lives. It reaches true equilibrium only in death. Up until then, it constantly needs to accumulate energy, not dissipate it. Doesn't this make entropy its enemy? To answer this question, we need to direct our attention to the next step up in our staircase.

Stair 3: Biology

". . . Then a Miracle Occurs!"

I won't be surprised if you've seen this cartoon many times before, but that's because the point it makes never ceases to be relevant:

"I think you should be more explicit here in step two."

In our case, "step two" is the transition from chemistry to biology—from inorganic matter to life itself. Life is a self-organized, emergent system that continually replenishes itself through an unimaginably complex accumulation of staggeringly interdependent processes. The more you learn about it, the more improbable it seems. There is simply no apparent reason for it to exist. If there ever was a miracle that needed explaining, it is this one: *What is life, and how in the world did it ever come to be?*

As a point of departure, let's keep in mind where we are on the infinite staircase. Every stair has some relation to life, and most of them are still to come. What we are looking for on this stair is the *transition* to life, not life fully realized. The latter will occur largely under the influence of natural selection, a topic we address in the following chapter.

In that context complexity will evolve as a function of organisms competing for resources under conditions of scarcity. The transition in this chapter, by contrast, is driven by entropy. Here complexity evolves as molecules interact to shed energy under conditions of abundance.

As we saw on the stairs of physics and chemistry, when matter gets energized, it seeks to dissipate that energy to reach a state of equilibrium. That's what cooling is all about. In support of this effort, complex molecules can take energy offline by storing it in their bonds, thereby hastening the cooling. Of course, if energy is scarce, then matter is not energized, and complexity does not emerge. That is why there is no life on Pluto. Similarly, if the energy is too intense, then there is no opportunity to cool down, so again complexity does not emerge. That is why there is no life on the sun. But if the environment is moderate, and especially if it oscillates between hot and cold the way Earth does when it alternates between night and day, or the way water does when it swirls around volcanic vents, then an energy gradient arises around which complex molecules can evolve. That is why there can be life on Earth.

But just because there *can* be life on Earth does not mean that there has to be. To account for that transition, there are two questions we must answer:

1. What are the necessary precursors for a transition to life?
2. Given these precursors, how could inorganic matter self-organize into persistent dynamic systems upon which natural selection could act?

We can start by inventorying the ingredients that must come together to enable life as we know it to emerge. These include:

- A plentiful supply of the subset of atoms that make up all living organisms—principally, *carbon, hydrogen, oxygen, nitrogen, phosphorus*, and *sulfur*.
- An abundant source of energy to drive the unfavorable chemical reactions needed to build complex molecules from these atoms—a precursor of *nutrients*.

- A container to make these molecules in, so the various piece parts don't wander off—a precursor of *cells.*
- An initial set of complex molecules to serve as building blocks that can be assembled into higher-order structures—specifically *amino acids, nucleic acids,* and *lipids.*

Assuming such an inventory can accumulate all in one place, we can then ask what processes must emerge to support persistent self-replicating systems upon which natural selection could act. These include processes for:

- Generating dynamic systems out of our building blocks—a precursor of *biological pathways.*
- Fabricating specified structures—a precursor of *RNA* and *ribosomes.*
- Specifying and controlling what structures are to be replicated and when—a precursor of *DNA.*
- Accommodating all its pathways as it expanded, and eventually reproducing itself to expand still further—a precursor of *cell membranes.*
- Transporting its energy system so that as it reproduces, it can expand into adjacent territory—a precursor of *metabolism.*

Over the course of this past century, science has developed a set of competing narratives about how these ingredients and processes might have come together to enable the emergence of life. Amino acids could have been created via lightning strikes. Complex molecules could have been organized by aligning them with the crystals in clay. Replication could have begun with the emergence of RNA as a precursor to DNA. Life could have emerged under deep layers of ice protecting it from the destructive impact of the sun's ultraviolet rays. All these theories entail the idea of self-organization arising spontaneously under the influence of entropy, so any one of them could be used to continue our story.

One of the most plausible of these narratives, the one we will trace here, is described by Nick Lane in *The Vital Question: Energy, Evolution, and the Origins of Complex Life.* In Lane's narrative, life began around 3.5 to 4 billion years ago—less than a billion years after Earth itself formed. Geographically, it developed under the sea—yes, Earth was a water world even way back then—specifically at *alkali hydrothermal vents*. These are places where spreading plate tectonics below expose underlying rock and gases to the seawater above, resulting in an upsurge of material, creating pumice-like microporous formations chock-full of the atomic elements needed to create life.

That provides a plentiful supply of our basic atomic ingredients. To create anything from them, however, requires energy. Virtually all life today gets its energy, directly or indirectly, from the sun, but in Lane's narrative the energy used to initiate life came not from the sky, but from the earth and sea. Specifically, alkali upflow from hydrothermal vents, rich in positively charged hydrogen ions, gives rise to electric currents as it moves adjacently to the more acidic, negatively charged seawater. This can happen, however, only if the two domains are kept separate enough to allow electricity to flow between them. In an alkali hydrothermal vent, this separation is provided by the very thin walls that make up its microporous rock formations. The result is the spontaneous formation of an electrical field that can be harnessed to do work.

Thanks to the serendipitous appearance of hydrothermal vents, we have all the atoms we need, we have an environmental energy source to drive unfavorable chemical reactions, and we have cell-like microporous containers to keep our work in process from drifting away. Now we need to figure out how our environmentally provided electric field can energize the creation of life's three basic building blocks—the amino acids that make up proteins, the nucleic acids that make up DNA and RNA, and the fatty lipids that make up cell membranes.

In Lane's scenario, biochemicals accumulate in the micropores of the vent in concentrations that enable them over time to self-organize into more complex molecules. The walls of these pores are "leaky," meaning that protons can readily pass through them. This creates the necessary electrical current to catalyze the binding of atoms into increasingly complex molecules. At the same time, however, the pores are not so leaky as to let the molecules themselves escape. As a result, as populations build, they remain available to serve as building blocks for higher levels of organization.

That complexity should emerge under these circumstances is not miraculous. Instead, it is *inevitable*. Why? Remember the principle we derived from our discussion of entropy. All matter is predisposed to configure itself to reach its lowest possible energy state. When atoms get all jazzed up, as they start to cool down, sometimes they get tangled up with one another and make molecules. As we saw from chemistry, that state involves atoms bonding with other atoms, releasing free energy in the process. So the resulting structure is both more complex and stable. Because it is stable, it can now participate as a component of an even more complex structure, which it is predisposed to do, presuming it helps that artifact get to its lowest possible energy state. What all this means, in essence, is that *whenever the universe can bind, it will bind*. Ever since the Big Bang, matter has been doing everything it can to cool down. In so doing, it will exploit every pathway to shed energy. Creating increasingly complex artifacts, it turns out, is an unintended consequence of this process.

Of course, life is not just complex—it is *organic*. Complex molecules, in other words, are just the beginning. For life as we know it, whole complexes of complex molecules have to interoperate in predictable ways and on a repeatable basis. This begins with the emergence of biological pathways.

Biological pathways evolve from random chemical reactions that become sustained through positive feedback loops. That is, while most random reactions lead nowhere, occasionally one of the outputs from one reaction will serve as an input to a subsequent reaction. And a fraction of those reactions will produce an output that will be an input to a third reaction, and so forth. Finally, a chain of such reactions can come full circle, such that an output from the last reaction serves as the input to the first reaction. Now we have a self-sustaining cycle—essentially, molecules interoperating in a predictable way on a repeatable basis. Whatever other outputs are generated during that cycle can be said to be produced by it. We have, in effect, a production facility, a mill wheel driven by the river of entropy.

What does it take to keep such a facility in operation? The short answer is the right combination of positive and negative feedback—positive enough to keep it from stopping, negative enough to keep it from spinning out of control. Tornadoes, flocks of birds, and ant trails all represent self-sustaining cycles that spin up, persist for a while, and then spin down. In each case, the result is an emergent entity that has its own distinct set of attributes. But unlike living organisms, none of these persists because each one, sooner or later, exhausts its source of positive feedback. Why didn't the same thing happen to those original biological pathways forming around the hydrothermal vents?

The answer is that the vents provided a natural energy gradient based on how far the pathway was located from the vent. Too close, and it can't settle down; too far, and it loses momentum. But in between there will be a Goldilocks point where, assuming the vent itself persists, a pathway can form and be continually fed with just the right amount of energy to keep it going. It is at such Goldilocks points that, according to Lane, the precursors of life got their first start.

Taking the story forward from here, however, is a challenge. There are innumerable combinations and permutations of biological pathways that enable the sustained existence of just a single cell—any cell. Take away any one of them, and the cell dies. But they could not have come into existence all at once. So while there must have been some kind of linear progression over time, it is impossible to determine it. And since life has evolved under conditions of natural selection for billions of years, it is equally impossible to say how much of what we have now would have existed then.

Here's what we can say, however. Biological subsystems that are shared across virtually all living things must have emerged in a relatively complete state very early on in life's trajectory and evolved only minimally thereafter. That is, driven by entropy, they came into existence once, and probably only once, subsequently conserving their form and function across eons of time despite all the other kinds of natural selection that ensued. They are, in effect, the primordial artifacts of life's emergence. To complete our story of the transition to life, we simply need to show how each of these core engines could have emerged in the first place.

The subsystems we have in mind are:

- a protein factory called the ribosome,
- proteins themselves,
- an information system consisting of DNA and RNA,
- cellular membranes,
- ATP-driven metabolism, and
- the eukaryotic cell.

Once this cast of characters is assembled, we can build a credible narrative around how natural selection enabled increasingly complex forms of life to evolve. But how did this cast get assembled in the first place?

Let's begin with the ribosome and the fabrication of complex molecules. How can molecules fabricate other molecules, and why would they want to? Well, the answer to the second question is our old standby: to shed excess energy in service to increasing entropy. But how do they actually do that?

To begin with, the vast majority of all the complex molecules fabricated in our cells are *proteins*. Forget everything you learned about diet—we are not talking food groups here. We are talking about a set of 20 different amino acids that connect up like children's pop-bead necklaces into chains of indefinite length.

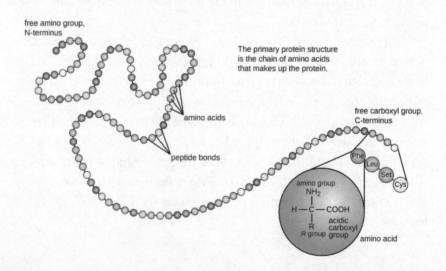

free amino group, N-terminus

The primary protein structure is the chain of amino acids that makes up the protein.

amino acids

peptide bonds

free carboxyl group, C-terminus

Phe
Leu
Ser
Cys

amino group NH₂

H—C—COOH

R group   acidic carboxyl group

R group   amino acid

Of the five hundred or so amino acids known to exist, there are exactly 20 that plug and play to create every one of the tens of thousands of proteins that enable life on our planet. The sequence of amino acids needed to fabricate each protein is coded in DNA and ferried to the manufacturing facility via RNA. All living things use this same set of 20 pop-beads, making them a perfect example of a primordial process.

Because of slight differences in electric charge, various bits in each chain can get attracted to various other bits, the result being that protein chains fold back on themselves to create unique three–dimensional shapes, something like a pile of spaghetti. Once this spaghetti pile finds its lowest energy state, it stabilizes and can be treated like a hardened pile of spaghetti. These are the building blocks with which life works.

As you can see from the illustration below, each self-assembled protein is filled with nooks and crannies and studded with little projections. All of these are characterized by a unique configuration of electrical charges. When a projection in one folded-up protein matches up with a nook or cranny in another, and just the right opposite charges line up to attract one another, then the two proteins connect, like a spaceship docking at a space station.

Now here comes the fabrication part. As they adjust to their new configuration, the distribution of electric charges across the two molecules will change slightly, causing one or both of the proteins to actually

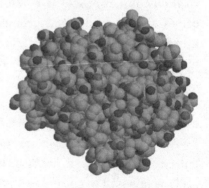

folded protein chain
(main chain view)

folded protein chain
('space-filling' view)

change their overall shape. That change in shape, in turn, can have further consequences. It could expel some other protein from some other docking station elsewhere on the protein, or it might open a new nook for some other protein to dock into. A cluster of such proteins interacting in tandem can—again, by pure accident—create a workbench in which a molecule under construction can be temporarily held in place while it is being assembled and then released when it is complete. It all occurs through the interaction of three-dimensional shapes changing configurations as different molecules are added or subtracted, all driven by the entropic need to find their least energized configuration.

The workbench at the heart of protein fabrication is called the *ribosome*. By poking it in various places, other molecules, acting as catalysts, cause its nooks and crannies to reconfigure in ways that accept a new molecule. Once the new molecule is bound in place—an event that further changes the configuration of nooks and crannies—other molecules can be attached to it, or it can be attached to other molecules, in a predictable way. This is what enables the ribosome to bind and chain amino acids into proteins.

Now, absent the other necessary subsystems to direct this effort, this equates to *blind fabrication*. That is, the ribosome is just randomly binding whatever it can, whenever it can, simply because creating complexity—any kind of complexity—furthers the dissipation of energy in accordance with the Second Law. By contrast, *managed fabrication* consists of selective binding to create specified complex molecules. Once established, through repetition it can then build up an inventory of complex molecules to be used as building blocks for even more complex entities—systems on top of systems, complexity on top of complexity.

To transition to such managed fabrication, however, two additional elements need to be incorporated into the replication system: DNA and RNA. These are the molecules that encode the specifications for proteins. Both, it turns out, are also chains, but they are made from nucleic acids as opposed to amino acids. DNA is the one you are most familiar with. It comes in a double strand that self-assembles into a helix, the shape of a spiral staircase. RNA, by contrast, comes in a single strand. Unlike proteins, however, which build up their chains from 20 different pop-beads, nucleic acids make do with only four. In the case of DNA, they are abbreviated as A, C, G, and T, and for RNA as A, C, G, and U. T and U are direct equivalents, meaning that either molecule can code for exactly the same sequence of amino acids as the other one. Thus, both sets can be used to encode the same set of instructions for programming the ribosome.

There is, however, an important difference between them. As a single strand, RNA can move about the cell much more easily than DNA. Also, it can crumple up on itself like a protein and thus do double duty, both as a two-dimensional transmitter of coded information and as a three-dimensional nook-filling catalyst for building actual proteins. Indeed, the ribosome itself is partly made up of proteins and partly of crumpled-up RNA. For these reasons, most scientists believe that RNA was first out of the gate in getting replicated life going.

There is a downside to RNA's structure, however, which creates the opening for DNA. When an RNA coding chain gets damaged or mutates in any way, it loses the information it was carrying for good. DNA, by contrast, has a built-in fidelity mechanism because each of its two helical strands is a perfect mirror image of the other. Every step in its spiral consists either of an A pairing up with a T, or a C pairing up with a G. So if one strand reads A, T, C, G, then the other side of the helix has to read T, A, G, C. Given these fixed pairs, if one of the bases gets damaged, or if the fabrication mismatches a pair, the cell can recover the information by reading the other strand—and astonishingly, that is exactly what the DNA repair mechanisms in your cells do millions of times every day.

In two-dimensional linear form, RNA and DNA operate just like computer code written out in a line of text. Instead of using bits made up of zeros and ones, however, they code messages using their four different bases. And instead of being processed in eight-bit bytes, they are grouped into triplets called codons. Because there are four bases being read in groups of three, there are a total of 64 uniquely different combinations to work with, reaching from AAA to TTT. With the exception of three codes that all represent the STOP instruction, each of the 61 remaining combinations code for one, and only one, amino acid.

In this context, a gene is simply a long string of codons that specifies the sequence of amino acids necessary to make up a particular protein.

This is how DNA provides the blueprint for building every possible protein in every living organism. Because there are a lot more combinations
(61) than there are amino acids (20), the system has built-in redundancy,
which makes coding a lot more fault tolerant since some errors in tran-
scription will lead to the same result.

To actually build a given protein, however, requires some help from
RNA. It comes in several forms. *Messenger RNA* copies the instructions
from the DNA and ferries them out of the nucleus, into the cytoplasm,
and over to the ribosome. There it pairs with a *transfer RNA* molecule
that serves as a matchmaker to bring the right amino acid to the workbench. Thus enabled, the ribosome fabricates the entire universe of protein molecules that make up the bulk of you and me. Of course, that
raises the question, What are all these molecules used for?

The short answer is a lot! Some proteins attach themselves to DNA
molecules to trigger the expression of one gene or inhibit the expression of another. That's what allows cells to differentiate in the first place.
Other proteins support metabolism by acting like pumps to push out
charged ions across the cell membrane so they can help drive the next
generation of reactions when they are let back in (thereby mimicking
the flow of protons from the alkali thermal vents). Some proteins form
into stiff molecular rods that serve as rails upon which to transport stuff
around inside the cell. Others act like trains on those rails to ferry molecules to their target destinations. Some act like poles to push against
the cell membrane to move the cell in a new direction. Some act like
ratchets to pull muscle cells across a substrate to create more leveraged
kinds of motion. Another class of proteins participates in signaling systems to communicate the internal state of the cell to entities outside it
and to receive their communications in return. Still others function like
flexible portholes that let certain atoms and molecules come into the cell
while keeping others out (thereby further mimicking the porous micro-
structure of the alkali thermal vent). In sum, proteins are the work-
horses of life.

Of course, for this work to unfold, it needs to have a home, a container, a cell. All life is cellular, with the earliest cell membranes most
likely formed from fat molecules called lipids. Lipids are molecules that
have one end that likes to hang out with water molecules (the hydrophilic

end) and one end that hates to (the hydrophobic end). As a result, they spontaneously align and group together in water to expose their hydrophilic side and protect their hydrophobic one. This leads to sheets of lipids (think oil films) as well as to spherical micelles (think bubbles). Micelle bubbles actually consist of two layers of lipids such that both their insides and outsides can be hydrophilic and thus be exposed to water, the hydrophobic ends being sandwiched safely inside the double layer in between. In this way cells keep an inside world of water separate and distinct from the outside world of water. This allows for different concentrations of ions to be inside and outside, thereby creating the electric fields that drive biological pathways and help transmit signals across nerve synapses. Similarly, lipids are used inside the cell to create compartments and organelles that isolate one kind of chemical reaction from another.

With cells in stock, there is just one more primordial artifact we need to account for the transition to life: a metabolic system that can create and store energy. This energy system is needed to activate and sustain the various chemical reactions necessary for life. Its emergence represents a singularity in its own right because it turns out that all living things use exactly the same system. Specifically, all metabolic systems feature a molecule called ADP (adenosine diphosphate), which, when energized, becomes ATP (the D for di- becomes a T for tri- by adding on another phosphate). This additional phosphate bond provides an energy source that can be used to power a wide variety of reactions. Whenever it is so used, the third phosphate pops off, and ATP reverts back to ADP.

To power life on a continuous basis, ATP must continually be re-created from ADP. This is accomplished by a complex molecule called ATP synthase. The speed and scale at which this is done is mind-boggling. Every cell in your body consumes on the order of 10 million molecules of ATP *per second!* You have tens of trillions of cells in your body. Add it all up, and that's a production quota equivalent to your entire body weight every day. But you only have about a pound or so of ADP around at any given time, so an ATP synthase molecule's work is never done.

The ADP-ATP energy subsystem is core to life's ability to sustain itself independently. Prior to its arrival, as long as proto-cells hung out in the pores of alkali hydrothermal vents, they could get their energy for free and thereby go about their complexifying activities. But they could not migrate beyond the vent itself, or even to the less active portions of it, without first developing some more portable metabolic system. That's what the ADP-ATP complex of molecules provides.

This energy system, combined with the DNA-RNA replication system, the ribosome-based fabrication system, and the lipid-based cellular containers, provide enough ingredients to invent life. As far as we can tell, life's invention was a singularity, meaning (1) it was not intended or planned and (2) it happened only once. That one event was enough to generate an enormous variety of single-celled microbes, mostly bacteria, and for around two billion years, that is all that life on Earth did. That gave cells plenty of time to get really, really good at what they do. Then yet another singularity happened.

Somehow a bacterium we now call a mitochondrion found a way to get inside another single-celled organism, called an archaeon, and the two commenced upon a relationship that biologists call *endosymbiosis*. The term refers to independent organisms existing in a symbiotic relationship with one living inside the other. This relationship truly did move heaven and earth.

For starters, mitochondria run exceptionally productive ATP synthase factories. By being able to multiply safely inside a single cell, they were able to raise the energy level of that cell by a factor of ten. For this to work, however, the cell had to evolve to cope with two different sets of DNA floating around each other. This led to the development of the cell nucleus, consisting initially of the archaea's genes, supplemented by a migration of many of the mitochondrial genes into that nucleus. At the same time, enough local genes remained inside the mitochondria for them to replicate independently. This enabled them to better adapt in real time to their own local challenges without having to go back to corporate to get authorization.

With the emergence of the nucleus and the incorporation of mitochondria, we get the first instance of the modern eukaryotic cell. This is the cell that makes up all complex organisms extant on Earth today, be they animal or vegetable. Its arrival inaugurates the beginning of evolution as we know it. Yes, the prior population of prokaryotic cells—essentially, all the bacteria—have continued to evolve in order to adapt to different habitats, but they are unable to drive increases in complexity because they lack the energy to do so. Eukaryotes changed the game.

One of the first big innovations eukaryotes introduced was eating. They were the first cells to engulf things. This allowed them to harvest energy that had been accumulated by another organism, getting it all in one gulp as it were. However, since there are only a limited number of other organisms to consume, this initiated a competition among organisms for scarce resources. That, of course, is the formula for natural selection, so it is at this point in life's history that evolution really takes off.

Under the influence of natural selection, some eukaryotes evolved ways of working together, creating multicellular populations. Later they evolved ways to differentiate cell populations within a single organism. This resulted from minute changes in protein bindings that expose different strands of DNA to replication, resulting in cells that develop into different tissues. Differentiated tissues, in turn, led to the emergence of organs, conferring whole new levels of specialized capabilities, thereby bettering the organism's ability to compete for scarce resources. Throughout this process, organisms were differentiating from each other, driven by the forces of natural selection to find niches in the ecosystem where they could thrive.

Somewhere along the line, a fateful branch occurred wherein a subset of eukaryotes performed a second act of endosymbiosis, engulfing chloroplasts. Chloroplasts represent a different kind of energy factory, one that uses sunlight as its source of energy, thereby giving rise to plants. Via photosynthesis, plants convert carbon dioxide and sunlight into sugar, oxygen, and water. As plant populations burgeoned, Earth's atmosphere became increasingly permeated with the oxygen being released by photosynthesis, and much of life converted from an anaerobic to an aerobic energy system, one able to generate and consume a much greater amount of energy than its predecessor. That led to the rise of animals that, among other things, ate plants, accessing their stored energy by simply engulfing it.

With the rise of plant-eating animals, we are far enough into the transition to life to step back and take stock of what has happened. In every case we saw that external sources of energy were activating levels of increasing complexity that in turn did their best to shed that energy to achieve their most stable state. All that energy that mitochondria gave to that original archaea cell—what was it going to do with it? It had to shed it somehow. Making more of itself as it always had would not get the job done; the energy backlog would keep building. It had to go up a level in complexity to consume more energy. That led to multicellular systems, which in turn led to tissues, then to organs, then to increasingly complex organisms, which in turn created populations, which in turn created ecosystems.

Each higher level of self-organization creates its own locus of energy consumption. Thus, organs not only consume energy themselves, they export molecules that serve as signals to other organs to alter their behavior to consume more energy as well. Organisms do the same in populations. Populations do the same in ecosystems. But how could such signals work? Not because a cell or an organ can "understand" them—cells and organs don't have brains. Rather it is because they trigger a molecular reaction that sheds excess energy, thereby aligning themselves with the one thing that the universe unceasingly wants to do—namely, chill.

Beneath all the complexity we see around us, in every case, atoms and molecules are simply reconfiguring themselves to achieve their least energetic state. At this level, that is all that is happening. Inevitably, we will talk about these phenomena using ideational language. For example, we say that DNA is a *program* that regulates our metabolism by *coding* for creating a specific protein. But molecules cannot think, nor can they code. All they can do is get excited by incoming sources of energy and then reconfigure themselves to shed as much of that excitement as possible. That's it. In other words, all the exquisite ordering of the universe—which is spectacular and miraculous beyond human comprehension—is at an atomic, molecular, and cellular level the result of a cosmic game of *jiggle the handle*.

How can that be true? How can one say that 23 chromosome pairs with 100 billion DNA bases coding for something as miraculous as your child, your spouse, or yourself—how could that really be the result of random jiggling? Well, to be fair, it is not *random* jiggling. It is *entropic* jiggling.

As we have been saying repeatedly, entropic jiggling is based on atoms seeking to transition to their least energized state. In the process, some of the atoms get tangled up with each other and create ordered structures by accident. Complex systems, in this context, are the unintended consequence of thermodynamics. Since Earth was formed, the sun has continually bombarded it with energy. Because our planet spins, any given patch of Earth oscillates between higher and lower energy states on a daily basis. Complex systems evolve under the stimulus of the energy gradients created by such a dynamic. Originally this happened underwater around hydrothermal vents, where the energy came from Earth's underlying volcanism, and the dynamic gradient was created by the distance away from the vent itself. But as cells became mobile and moved out into the open water and air, the sun became the primary energy source that drove development of further organic complexity. Our rising and setting sun provides a perpetually renewing energy gradient, and life transforms this fluctuation of energy into an increasingly elaborate set of cascading interactions.

It is not clear whether life had to emerge. Indeed, it is not clear whether it has ever emerged anywhere else in the universe. But once it did emerge, then nature—more specifically, the Second Law of Thermodynamics— began to select for complexity. Each gain in complexity allowed for more shedding of the free energy provided by the daily oscillations of a solar energy gradient, all operating in accordance with the Second Law. Or, as Atkins puts it,

> We all live off the spontaneous disposition of the Sun's energy,
> and as we live so we spread disorder in our surroundings.

Escalations in complexity beyond the molecular biology of the cell are driven primarily by natural selection operating on populations, or what we will call the metaphysics of Darwinism. That is the subject of the next chapter. But before we turn to it, we must answer a critical question:

As we trace the spontaneous emergence of complexity in organic systems, what connects the Second Law of Thermodynamics, the principle that governs the metaphysics of entropy, to natural selection, the principle that governs the metaphysics of Darwinism? We need to secure this connection if we are going to end up with a single, integrated staircase.

That connection comes in two parts. The first part is that entropy, as it turns out, is an indirect measure of complexity. It is complexity's shadow. That is, the more complex a dynamic system is, the more energy it consumes, and the more entropy it generates. The Second Law does not mandate the emergence of complexity; it enables it. Without the unceasing flow of energy from a location of heat to a location of coolness, there is no way that complexity could emerge. The more complexity created, the more energy consumed; the more energy consumed, the more entropy generated.

The second part of the connection is that natural selection also favors the emergence of more complex systems. Why? Because they can employ a greater variety of strategies for living than simpler systems. This makes them more competitive whenever metabolic resources become scarce. Thus, for starters, more complex organisms can simply consume simpler ones. The first organism that "fed" on another organism fueled its own entropy-creating metabolism. By virtue of the Second Law, we know that some of this fuel was not utilized but was instead released into the environment as heat. At the same time, this process also deconstructed a previously well-ordered system—namely, the body of its prey. Wherever order becomes disordered, matter reverts to a simpler, more stable state, and once again, entropy increases, again with a release of energy into the environment.

As a corollary to this connection, we can assert that the amount of entropy a living system generates is an indirect measure of its complexity, and we can rank-order systems in terms of complexity by comparing the amounts of entropy they generate. In this context, the very act of eating, by itself, puts animals higher up in the entropy generation hierarchy than plants, and it positions mobile animals higher than anchored ones because they can seek out their prey rather than having to wait for it to come to them. And bigger is better, too, since in general, the bigger the animal, the more cells it has, the more energy it metabolizes, the more it needs to eat, and the more entropy it creates. Size, of course, is normally covariant with population. That is, a few bigger animals is one way to achieve a local maximum, and a greater number of smaller animals is another. For this reason, the transition from solitary individuals to populations is another important ratchet point in the evolution of complexity.

In populations, collaborative synergies allow organisms to create more entropy acting as a group than they could by simply summing up their independent operations. We see this largely in the domain of environmental impact. All animals (as well as plants and fungi) have some capacity to reshape their environments to create niche habitats more favorable to their particular species. There are all kinds of wonderful examples of such niche management—from networks or tree roots and fungi, to anthills and beehives, to bird nests and beaver dams. All these populations thrive by modifying their environments in ways that increase entropy. So now we have a "three-fer" in the entropy creation game: (1) metabolism *plus* (2) consumption *plus* (3) environmental impact.

Note that what we are building here is a kind of Great Chain of Entropy. It aligns directly with its predecessor, the Great Chain of Being. Both are organized by levels of increasing complexity. And at this point we are ready for the truly dramatic transition to humans. From a metabolism and consumption point of view, we humans are not particularly distinct from our mammalian cousins. But from an environmental impact point of view, we perform so spectacularly as to leave even our closest relatives far behind. As a species, we are Earth's foremost generators of entropy by multiple orders of magnitude.

This may seem odd at first because we still tend to think of entropy as a force of disorder and of ourselves as continually combating entropy in service to creating and maintaining order. But as we have been observing repeatedly, entropy is the exhaust we must inevitably create whenever we bring order into existence. And no matter how much order we create, we must generate even more entropy, or else we violate the Second Law.

Consider Michelangelo's *David*. It is hard to imagine a more exquisitely ordered artifact—where is all the entropy? It went into the surrounding environment. From the time that the marble was quarried to the hauling of the marble to Florence, the making of the wagon to haul it, the breeding, raising, and feeding of the oxen and the men to pull it, the chiseling away of all the marble that was not David, the sanding, the polishing—every single step transformed energy into order and in the process released even more energy to entropy. And that's just to make the *David*. Now make all of Florence. Now make all the cities of Italy. Now make all the cities across the whole world. Then add the factories, the pipelines, the railroads, the power plants, and you begin to get the point. When it comes to creating entropy, human beings are in a class all by themselves.

To pursue this entropy-generating cascade of integrated activities further we must proceed to higher steps in our staircase. That will be the focus of the following two chapters. Inevitably they will take us farther and farther away from the domains of physics, chemistry, and biology, but we must not lose connection with these lower stairs.

To be sure, the metaphysics of entropy is grounded in materialism whereas much of what we will discuss going forward is ideational rather than material. Ever since Descartes, Western philosophy has seen these two domains as antithetical, but that is simply a mistake. There is no thought without matter. There is no mind without brain. From the universe's perspective, each subsequent step in the staircase is a natural extension of the principles of materialism because every step unfolds according to the Second Law. At the same time, however, materialism must now be left behind because higher orders of complexity embody novel emergent properties that it cannot explain. Any attempt to reduce social and intellectual developments to materialism, to reduce thought to matter, to reduce mind to brain, must fail. It is not that materialism in and of itself is incompatible with the *David*—it just cannot account for its most interesting properties.

That's why we are heading for higher ground.

CHAPTER 3

The Metaphysics of Darwinism

Before we enter this second stage of our journey up the staircase, let us remind ourselves what we mean by *metaphysics*. It's pretty simple, really. Metaphysics is our best explanation of the situation we are in, of why everything exists and how everything happens. In that context, the prior chapter did its best to explain material reality—how it works and how it came to be. This chapter will build a bridge between that world of mindless matter—the stairsteps of physics, chemistry, and biology—to a world of matterless mind: the stairsteps of language, narrative, analytics, and theory.

This transition is enabled by the emergence of consciousness. The development of consciousness is triggered by the emergence of desire, and it gives rise to the emergence of both values and culture. Importantly, all of this happens prior to the emergence of language. Thus, consciousness is grounded in materiality and animality before it connects us to our higher faculties.

The Infinite Staircase
(The Metaphyics of Darwinism)

11. Theory

10. Analytics

9. Narrative

8. Language

7. Culture

6. Values

5. Consciousness

4. Desire

3. Biology

2. Chemistry

1. Physics

Emergence in this section of the staircase is driven by the two great shaping forces of evolution: natural selection and sexual selection. Both were described brilliantly by Charles Darwin. Prior to Darwin, it was impossible to understand how complex life and human culture could have come to be without an intelligent designer imposing an intelligent design. So what exactly did Darwin say that changed our perspective so drastically?

What Is Darwinism?

The root metaphor for understanding Darwinism is *selective breeding*, as exemplified by the raising of cattle, racehorses, and pedigreed dogs, or the cross-pollination of crops or of ornamental flowers. Breeders in every case select from each generation those members that best exemplify an idealized set of traits and then mate one to another with the expectation that the subsequent generation will exemplify the ideal even more closely. Such breeding represents *intentional selection* in service to *intelligent design*.

Natural selection proceeds by an analogous mechanism, but it consists of *unintentional selection* that leads to *emergent design*. This occurs spontaneously wherever living beings compete for scarce resources or strive to survive under adversarial conditions. Winners in these compe-titions will sometimes possess traits that make them more competitive. Because these winners are more likely to survive to create the next gen-eration, their traits get an increasingly disproportional representation in the overall population. That increases the pool of genes that code for them (although no one in Darwin's time had any knowledge of genetics). Losers, by contrast, either do not survive or do not thrive. Either way, their contribution to the gene pool diminishes over time. As a result, populations as a whole become increasingly more effective and efficient at defending themselves and securing the scarce resources they need to survive. They succeed because they employ better strategies for living.

Natural selection, however, represents only one of the two critical variables that lead to evolution. The other is sexual selection, which is particularly influential among mammals. It involves finding a com-patible and cooperative mate. Sexual selection is normally led by the female of the species. She leads because, in the majority of cases, she has a significantly bigger metabolic investment at stake in reproduction than does the male. What traits she selects for can sometimes seem, at least to males, quite arbitrary. There is reasonable speculation, however, that such traits ultimately do correlate with better mate performance, in either the production of offspring or their subsequent care. In any event, sexually attractive males mating with discriminatingly selective females end up transmitting their genes to the next generation, which is again subjected to the same forces of natural and sexual selection.

The result of natural and sexual selection is increasing specializa-tion within species and increasing variation among them. All this hap-pens without any input from an intelligent designer. But what drives this process in the first place? What makes organisms fight to survive and mate? To answer that question, we need to hitch a ride on a streetcar named Desire.

Stair 4: Desire

The Darwinian Mean

Desire is the motor that drives all animal behavior, including our own. It arrives on the evolutionary scene prior to consciousness, and consciousness itself develops directly under its influence. Understanding its foundational position and role in the infinite staircase is critical to putting the levels above it in proper perspective.

Once again, we need to be cognizant of where we are on the staircase. As humans, we experience desire as embedded with all kinds of cognitive content. But prior to consciousness, desire consists of mindless motivation. Take hunger for example. Simple organisms like flatworms or single-celled amoebas do not experience hunger—they have nothing to experience it with—and yet they behave as if they did. What they are actually doing is reacting either to biochemical signals generated by their own metabolism in response to falling levels of nutrients inside their cells, or to external signals transmitted via receptors in their cell wall signaling the presence of nutrients nearby. Either type of signal galvanizes the organism to seek food to fuel its continued existence. Once nutrients are secured, then the behavior ceases, again, as the result of other chemical signals. Overall, the organism is being governed by positive and negative feedback, resulting in the dynamic equilibrium we call life.

Biochemical feedback systems are fundamental to all living things, beginning with plants. Plants have no nervous systems, but they do have vascular systems that can carry chemical messages back and forth. Different chemical concentrations, arising from changes inside the plant itself or in the environment, stimulate responses that change the configuration of leaves or the growth of stems or the extension of roots. Natural selection can act on these changes such that over time populations of plants become increasingly adept at adapting and reproducing. Sunflowers evolve to follow the sun, legumes to have symbiotic relations with microbes to access life-enabling nitrogen, and fruit trees to exude scents to attract the insects and birds vital to their pollination. All these behaviors are strategic and therefore can look intentional, but of course there is no mind involved at all.

In animals, this same kind of chemical signaling is also at work as various internal organs message each other via their circulatory systems:

- We need some more oxygen over here!
- Hey, I'm bleeding!
- Boy, am I full!

These signals in turn trigger responses in other subsystems, causing them to emit other chemicals that signal further changes, and so on. Over time, a kind of biological economy develops in which subsystems learn to scratch each other's backs. Needs can be signaled and fulfilled, and availability can be announced and pursued. We call these signals *hormones* if they circulate internally and *pheromones* if they circulate externally. Since pheromones activate hormones, in the end it is all about the latter. A life under the influence of desire is nothing if not hormonal.

In animals, however, this back-and-forth signaling takes on a whole new dimension with the emergence of nervous systems. Now messages transmit virtually instantaneously. The motivations are still chemical—it is just that the time to react to them has been taken effectively to zero. Such low latency allows for much more complex behaviors to emerge, beginning with mobility. These behaviors are coordinated via centralized nodes in the nervous system, structures that will evolve to support consciousness. At this point in the staircase, however, they enable a transitional state of complex reflex actions we call instinct. Instinctual behavior is strategic, and is thus shaped by natural selection, but it is not yet conscious—think of ants building anthills, spiders spinning webs, or birds flying in flocks. That said, this behavior is driven by desire.

Desire in its most generally applicable sense is the innate preference of all organisms to pursue their own well-being. The technical term for such well-being is *homeostasis*. Pleasure is the positive feedback we get when we are succeeding in our pursuit; pain, the negative feedback that we are not. Fear and anger are responses to increasing negative feedback, just as happiness and joy are to increasing positive feedback. These are core motivations that drive behavior and shape our strategies for living.

As physical beings, we engage in homeostasis all the time. We sit down and cross our legs. We scratch our nose. We take a sip of water. We get up to go to the bathroom. We adjust the thermostat. All these actions are in service to getting us back to a state of equilibrium that makes us feel better. The same holds true for the stream of random thoughts that come to mind. *I should answer that email. The car needs washing. Did I turn off the stove? How long until dinner?* Again, this is just the mind seeking to get to its own place of peace—a kind of mental homeostasis, in which unconscious desires are working themselves out through involuntary reflex behaviors.

The homeostatic state we are seeking to maintain is a kind of *Darwinian mean*. It consists of striking a balance between playing offense and playing defense, between proactively pursuing pleasure, happiness, and joy, and reactively responding to pain, fear, and anger. Unlike the more familiar Aristotelian mean, it is established prior to consciousness and language. It is not, therefore, a philosophical mean based on rational analysis; it is a biological mean based on a balance of hormone flows that "feels right." Neither the flows nor the feelings in this balance are chosen or defined by the conscious mind. Rather, they represent a population of possibilities driving a variety of behaviors against which various forces of selection can operate. The tendencies that result are behavioral norms that represent a given individual's strategy for living and, in a larger context, the social norms that represent a given community's strategy as well.

This has far-reaching implications for everything that follows. The higher levels in the infinite staircase inherit their tone and tendencies from the Darwinian mean of the communities from which they arise. These communities do not dictate our individual norms—there is still plenty of freedom for choice—but they do color every choice we make. Whether we embrace, repudiate, or modify these norms, one thing is for sure: we cannot do without norms of one kind or another. Homeostasis, by definition, is a return to a norm, and we are all wired to pursue it, like it or not. Our role and our responsibility as individuals is to choose, based on our own nature and circumstances, where to place the fulcrum of balance. There is no one size that fits all. Unlike the Aristotelian mean, in other words, the Darwinian mean does not converge toward an ideal. It is inherently relativistic and situational, and we need to respect it as such.

Specifically, this means we need to stay humble. Any time we advocate for a behavior or value, we need to contextualize it within the kind of situation that warrants it. All such situations are local. There is no such thing as a global situation, at least not until the sun or an asteroid has something to say about it. What we have instead are innumerable ways to create localities. That is, we can construct our own communities, build and fortify our own niches, and from such positions advocate for our situation and promote our values to others. In so doing, however, we must never divorce *prescription* from *description*. We must always acknowledge that while a situation does not dictate one's values, it inevitably grounds one's actions.

One universal element in our human situation is that we are all part of the animal kingdom. That means we are grounded in hormonal systems that cause us both to venture out and to fight back. We are driven by what one wag has called the "Four F's": feeding, fighting, fleeing, and reproducing. Every community has developed a homeostatic balance among these motives, one that serves as its social norm. Although these underlying forces are unconscious, the way we engage with them, how they shape us and how we shape them, is largely defined consciously. And with that thought in mind, we can now move on to the next step in our staircase—to consciousness itself.

Stair 5: Consciousness

A Darwinian Theory of Forms

Contemplating the emergence of consciousness can be tricky because we tend to use our personal experience as a point of reference. This is a mistake. Our personal consciousness is deeply immersed in the higher levels of the staircase, including language, narrative, and analytics. But consciousness itself arrived on the scene eons earlier than these distinctively human capabilities. To get the emergence of consciousness properly in perspective, we need to focus on animals instead.

When we turn our attention to animals, and particularly those of a higher order, we witness the operation of desire giving rise to strategic behavior of all kinds. Ducks beg, crows steal, crocodiles lurk, deer graze, peacocks preen, lions hunt, antelopes herd, egrets stalk, robins nest. What all these behaviors show is that *attention* is operating in service to *intention*. Consciousness, in other words, comes into being with an agenda. Its function is to deliver strategic outcomes in dynamic situations, all in service to desire.

Let us consider, then, what animal consciousness would have to include in order to perform the strategic behaviors upon which natural selection could act to drive evolution. Here is a first cut at such a list:

- *Awareness*. It has to register the organism's current state, specifically in relation to desire and other hormonal conditions. Without awareness, there is no starting point.
- *Intention*. It has to direct itself to perform actions that address its hormonal needs. Without intention, it cannot perform as an agent.
- *Perception*. It has to register the environment—specifically, presence and movement. Without perception, an agent cannot engage with any forces outside itself.
- *Association*. It has to connect its perception of things in its environment with the ability to change its current state. Without association, engagement is random.

- *Memory.* It has to remember both its actions and the outcomes they generated. Without memory, again, engagement is random.
- *Learning.* It has to get better as it gains more experience. Without learning, there is no increase in competence, and thus competitiveness.

This is quite a list, and it is by no means certain we have exhausted all the items required. What we can be sure of, however, is that we built this list based on a single premise: consciousness emerged because it enhances the competitive capabilities of organisms seeking to realize their desires. This implies two principles that are fundamental to all that follows:

1. *Consciousness is inherently strategic.* It is fundamentally in service to realizing desires, and all its properties are best understood in this context.
2. *Strategies for living are its core stock and trade.* The quality and effectiveness of these strategies are systematically improved through natural selection. This is how and why evolution favors the development of higher levels of consciousness.

The credibility of these two claims is based on equating strategies for living with *mental patterns.* Once we reach the level of language in the staircase, we will call such patterns *ideas.* In so doing we will inextricably bind them to words. But prior to the emergence of language, it is not clear what to call them. Can animals have ideas? If not, then what is going on in their minds, and how can our concepts of consciousness and strategy incorporate it?

In human experience, the closest analog we have to nonverbal animal consciousness is athletic performance. Clearly it is a conscious activity, but just as clearly, at least when performed at a high level, it is a nonverbal activity. One can, of course, engage with it verbally, thereby critiquing performance with a goal toward improving it, but it is not necessary to do so. Indeed, there is good evidence that analytic thinking actually disrupts the flow of athletic performance in the moment. This suggests there are two streams of consciousness operating to some degree independently, with many gifted athletes focusing on *feel* more than on *analysis*. Feeling, in this context, is a pattern-detecting faculty. It is also closely linked to homeostatic responses to hormonal rewards or the lack thereof. As such, we can use it to model how consciousness emerges under the influence of the Darwinian mean.

Such a model would help resolve a question that has held the Western intellectual tradition hostage ever since Descartes: Are the ideas we have about the world actually real? This anxiety arises from a dualistic division of mind from matter, one that is a natural outgrowth of grounding our understanding of consciousness in language. Linguistically grounded speculation about consciousness leads to the suspicion that our ideas, our patterns, are just words inside our heads, that they do not necessarily reflect actual circumstances, that we are trapped forever inside a linguistic mesh unable to make reliable contact with the world outside.

But animal consciousness rebuts this line of thinking directly. If the mental patterns in our heads did not reflect actual forces operating in the world, then strategies based on such patterns would simply fail, and the organisms pursuing them would be eliminated through natural selection. Simply put, if the patterns did not exist, then neither would consciousness. There would be nothing useful to select for. Thus Darwinism dispatches dualism.

Ironically, Darwinism actually aligns with Platonism. That is, it too has a theory of forms. To be sure, its forms are not Platonic. They are not timeless, nor are they ideal, nor do they precede material existence. Instead of being idealistic, they are pragmatic. The forms that animal consciousness processes convey strategic advantage. Animals acting on these forms achieve competitive outcomes. Their survival testifies to that fact that these forms are directly connected to the material world and reflect its dynamics with sufficient accuracy to be useful.

Thus, consciousness, that mental state that we perceive as making us uniquely human, actually makes us animals first. As adults, we experience and analyze consciousness through the medium of language, but clearly animals, not to mention very young children, do not. Yet the behavior of both animals and children demonstrates all the functions we listed as making up consciousness. That is the basis for the claim that consciousness precedes both language and culture; indeed, it is prerequisite for them. To be sure, culture and language shape consciousness, but they do so retroactively. Their impact on consciousness is secondary, whereas consciousness's impact on them is primary. Our minds do not come from above. They come from below. Emergence works from the bottom up, not from the top down.

To understand why, we can compare animal learning with artificial intelligence (AI). AI clearly comes from above. It is a magnificent accomplishment, a product of the highest levels in our staircase. At present, massive increases in computational capabilities are pushing a set of frontiers in AI and machine learning far beyond the capabilities of any animal, human, or group of humans. When IBM's Deep Blue defeated world champion Garry Kasparov at chess, that was one defining moment. When Google successfully exposed the entire World Wide Web to direct search by untrained individuals, that was another. When the NSA surveillance algorithms ferret out a terrorist cell, when Amazon shows an uncanny knack for promoting the right offer at the right time, when Apple's Siri finds you the closest Starbucks, when cars from Google, Tesla, and Uber are self-driving, all are leveraging AI and machine learning. And frankly, this is just the tip of the iceberg. AI and machine learning are transforming our world in ways we still have yet to grasp.

Given the ever-increasing mismatch in scale and speed that favors AI and machine learning, what can animal consciousness bring to the table to counterbalance such advantages? In a word, *desire*. Desire drives intention, and intention directs animal consciousness to strategic ends *spontaneously*. Intention is internal to and intrinsic to animals; it is external to and extrinsic to machines. As a result, a machine cannot distinguish between strategic and nonstrategic outcomes intuitively. It must instead be given a set of measurable parameters by which to do so. Thus, it can be programmed to act in strategic ways, but only up to the limits of its programming. When exposed to genuine novelty, it has no necessity to create a strategic response.

Animal consciousness, by contrast, will always generate a creative response, one that spontaneously emerges from its role in pursuing the Darwinian mean. Whether that response is successful will be tested ultimately by natural selection, but a response will be made. Moreover, the greater the stress in the environment, the more active the hormonal reaction, the more unpredictable the strategic response, and the greater the variation of trials upon which natural selection can work. The entire system is biased to innovate under stress, to maximize the chances of finding a successful strategic response to a novel circumstance in a remarkably efficient way.

So chalk up one for animal consciousness. But of course, this raises another question: Couldn't one program desire into a computer system? Or couldn't the system simply "machine learn" desire through recursively imitating human responses? I think the answer to both questions is probably yes, although I find that prospect to be somewhat scary. If and when a computer is wired to compete for resources beneficial to it, eventually one could imagine it competing with humans for such resources either directly or indirectly. Today that competition feels containable, but as our society evolves more and more toward an intelligent Internet of Things, containment becomes increasingly compromised. That raises the prospect of having to negotiate with machines—where would we find the shared values? It is not clear how soon we will have to answer this question. Thought leaders as diverse as Bill Gates, Elon Musk, and Stephen Hawking have all raised the alarm, but there is an awful lot of ground to cover before we reach this point. Nonetheless, the fact that we can envision such a future state does tee up the transition to our next topic, *values*.

Before leaving the domain of consciousness, however, I want to address a topic of great importance to all humans, one that cannot be explored without reference to the world of language and narrative— namely, *our experience of self*. This requires us to temporarily leap ahead in our staircase framework in order to address that very private awareness we have of you being you or me being me, an experience that sets us distinctively apart from our fellow mammals. We all share this intensely direct experience of being alive, curious, engaged, committed—and it is not clear how it can be reconciled with a materialist/pragmatist view of the world. How does a world that evolves in complexity through purely mechanical means, with no outside agency of any kind, ever come up with anything as intriguing and compelling as *us*? If there ever was an argument for a divine creator, this is it. What credible explanation can we present as an alternative?

To address what makes human experience unique, we have to take a moment first to appreciate our brain. It is arguably the most complex thing on Earth. It contains around 100 billion neurons, the vast majority of which are present right from birth. Despite this mother lode of brain cells, however, the infant brain doesn't process well at birth because it lacks a sufficient population of synapses—the connections that link one neuron with its fellows. It turns out any given neuron can connect up with literally thousands of other neurons, so the possible universe of synapses is an extraordinary 100 trillion connections (that's 100,000,000,000,000). Of course, to build up that kind of a Rolodex, the brain has to develop at lightning speed, so perhaps it will come as no surprise to you that infants form as many as two million new synapses *per second* (although obviously that continues to astound me). In short, we are talking about one heck of a power source.

Now, what are these emerging synapses all about, and where do they come from? It starts with your five senses. Each sense consists of a network of nerves that, at one end, is connected to some kind of a detector (your eyes detect photons, your ears detect sound waves, your nose and tongue detect molecules in the air or in your mouth, your skin detects pressure and temperature) and at the other end, after some journeying, is connected to multiple parts of your brain. But here is the key point: as different as each of your senses are, by the time their signals get to the brain, they are all simply electrochemical currents. That's it. That's all your brain ever gets to experience. It can't see, it can't hear, it can't sense anything directly—it is sealed up in a dark and airless vat called your skull. Everything it knows, everything that happens inside your head, everything you experience, is just biochemically created voltage and amperage—period.

So, given no other context when we are born into this world, how do our brains learn how to make sense of anything? All they are fed are just electrical signals—basically noise—that activate some neurons that in turn pass the noise onto other neurons in an ever-increasing cascade across a massive galaxy of connections. What sense is there in that? And even if one could imagine making sense of this noise, who is going to be doing the sense-making? As Gertrude Stein once said of Oakland, California, there is no *there* there. How can things make sense if we are not present to make sense of them?

Here we can take a clue from the world of machine learning, specifically from a discipline called *deep neural networks.* Now, as you should know by now, I am no computer scientist, so bear with me as I do my best to represent what I believe to be the operating principles of a discipline I understand only by description. The question we are trying to answer, the one that parallels how the infant brain begins to make sense of the world, is how does a machine learning algorithm, with no training from any outside agency, process an undefined stream of data—essentially noise—and detect pattern within it?

The answer, it turns out, is *one layer at a time.* Suppose you expose a deep neural network algorithm to a data file consisting of, say, ten thousand cat pictures—basically, a whole lot of digital noise. Its programming is designed to detect repeatable regularities within this universe of signals, anything that could qualify as a pattern. Perhaps it begins by detecting that some pixels are light, and others are dark, labeling each as such. This creates a new representation of the noise, one that is sorted into light and dark. Call that Layer 1. Now, it takes this output, Layer 1, and processes it as input to Layer 2, to detect what additional repeatable regularities it might contain. Perhaps it detects that there are lines that divide light patches from dark patches. Take that output from Layer 2 and use it for input to Layer 3. That might lead to detecting the patches themselves—that there are zones of similar color and intensity, some of which are bigger than others. Now we have an algorithm that, by filtering the noise through Layers 1, 2, and 3, can detect shapes. Each subsequent layer inherits the learning of the prior layer and builds its classifications on top of it. Slowly, we are building an increasingly articulate map of the data we are being fed, one that will end up being able to identify cat photos on Facebook.

That's what machine learning does. Importantly, it is the sort of stuff that babies do, too. The early days and weeks of infant life are largely taken up with building deep neural networks to begin to make sense of the deluge of stimuli that accompanies their exit from the womb (where, by the way, they were getting their first exposure to electrochemical signals and already beginning to form their deep neural networks). All of this is made possible by a massive amount of computing power for processing signals, detecting regularities, and sorting data. As we have already noted, we are born with that computing power. It is what the brain with its close to 100 billion neurons can do from birth.

The infant brain does not know anything, but it can process signals, and thus it can begin to detect patterns. Which patterns stick? The most obvious choice would be those things that both repeat themselves and are associated with hormonal rewards. Akin to ants traveling back and forth on a path from their anthill to a source of food, similar nerve signals take similar pathways through the cascade of synapses, leaving similar chemical signatures in their wake. The result is a strengthening of those pathways, effectively remembering them. The more the same pathway is used—this is where the hormonal rewards come in—the stronger the memory becomes. Or as one scientist memorably put it, *neurons that fire together, wire together.*

So it is that newborns learn about their own bodies and learn about the world. In most other animals, much of this circuitry is actually prewired from birth as instinct, thereby enabling newborn horses or deer, for example, to learn to walk in a day instead of a year. We humans, by contrast, come into the world with much less preprogramming, which is why infancy lasts so long. There is just so much to learn, in particular because we use a big chunk of those first few years mastering language.

Language, as we shall confirm when we get to that stair in our framework, is a game changer. Once again, we can look to machine learning for a model of how our brains actually learn it. At the beginning, words are just a novel kind of noise, but our parents and siblings give so much attention to them that we quickly learn they are somehow important. So again, through a pattern of repetition and layering, reinforced by hormonal rewards, we are able to learn that words correspond to things and actions and properties. We also learn that they give us power over the world, as we can ask for things, say how we feel, make others smile. Finally, what was learned at first as an instrument for manipulating and responding to the environment becomes over time a vehicle for experiencing narrative and, later on, for creating narratives of our own.

Narrative is the key to understanding how we experience ourselves. We make sense of the world through stories of cause and effect, where characters play the part of various forces, and incidents and plots show the consequences of actions and relationships. From listening to our parents telling stories about us, we learn that we ourselves are characters in other people's stories, and eventually we tell such stories about ourselves. Finally, the moment comes when we begin to tell stories about ourselves to ourselves. We begin to narrate our own lives, silently, the way Snoopy does in the *Peanuts* comic strip. And like Snoopy, we insert commentary, make asides, and draw conclusions. Over time we build up a repertoire of memories, preferences, attitudes, and capabilities that we call *myself.*

Without language and without narrative there can be no self. Self-awareness—becoming self-conscious—is the realization that each of us is a unique character in a set of developing narratives that is organized around ourselves, a set of narratives that collectively we call *me.* The more attention we pay to this very interesting character called me, the more wired our synapses are to return to it as our point of reference and departure, and the more deeply it becomes entwined in our experience of the world. It is from this entanglement that all qualia emerge.

Qualia is the philosopher's word for the myriad of experiences that we register as *personal* or *subjective.* The smell of coffee, the mysterious glow of headlights in the fog, the taste of curry, the sound of our stomach growling—these are all qualia. We understand that other people have these same experiences, but we also know that no two people experience qualia in exactly the same way. What is going on here?

Qualia are best understood as the animal brain's equivalent of ideas. They are mental patterns that provide a mechanism for organizing sensations and memories in service to a desire-driven strategy for living. As our own experience testifies, we use these mental states to detect changes in our environment, be they pleasurable or painful, opportunities or threats. With respect to the emergence of human consciousness, qualia represent the infant brain's way of organizing experience prior to having language or access to narrative. This mechanism will stay with us for the rest of our lives, actively developing as our sensibilities grow and mature. Qualia are at the core of all esthetic experience. As such, they remain elusively out of reach of our language-dominated analytical capabilities. We want them to be subsumed by our linguistic worldview when in fact they antecede it. The developmental pathway is desire → noise → signal → response → fulfillment → qualia → language. Qualia will always be closer to our hearts than words.

Our subjective engagement with qualia is what gives us our sense of being alive. It's what gives us our sense of being important—important enough to have been *created*. That's one of the reasons we are so readily attracted to the idea of a divine creator. Well, it turns out, of course we are created. It's just that we are created from the bottom up rather than from the top down. In its own way this is an equally miraculous journey—arguably, even more so. At any rate, it is the journey we have set out on, taking the staircase one step at a time, from bottom to top. In that context, the past few pages have been a detour of sorts, leaping ahead to invoke the realms of narrative and language. It is time we got back to our step-by-step approach. Having seen how animal consciousness emerges from desire, now it is time to see how values and then culture emerge from animal consciousness, both of which occur before language appears on the scene.

Stair 6: Values

A Darwinian Social Contract

Values emerge as an external counterbalance to the internal urgings of desire. Both ends of this tug-of-war make claims on consciousness and, more specifically, on intention and choice. Values further the ends of the group, desire those of the individual. By virtue of this counterbalance, freedom of choice comes into being. This generates a range of possible strategies and actions upon which natural selection can act, at both the individual and group levels.

It may seem surprising to claim that values emerge prior to culture and language. In fact, however, they grow naturally out of the interaction of any conscious being with its social group. The former brings intentions driven by desires and fears; the latter provides boundaries and direction for sanctioned behaviors. Values, in other words, are socially constructed. Without social interactions, there can be no values, only desires and fears.

As a consequence, values are specific to social animals—primarily mammals. They do not emerge among reptiles. This initiates a fateful branching in the evolution of life on the staircase, with everything higher up being an extension primarily of the mammalian branch. Reptiles, by contrast, are driven solely by desire and fear. They are, from a mammalian point of view, inherently selfish and never trustworthy. It is no accident that our science fiction films portray those aliens who terrify us as having reptilian features, while those who align with us get a more mammalian look.

It is important to note that our deep connection to mammals is not due to us domesticating them and thereby imposing our values onto their behavior. Nor is it due to us anthropomorphically imposing our theory of mind onto their observed actions. Rather it is because we share common values that are mammalian by origin and only subsequently made human by refinement.

Two fundamental social interactions, both inherent in mammalian life, form the crucible in which shared values are formed. The first is nurturing the young; the second, governing the group. Both domains are controlled by socially empowered individuals—the mother and the leader—each of whom models and imposes behavioral norms. These norms are reinforced by hormonal rewards and penalties, and by this mechanism they become internalized and refined.

Turning first to the domain of nurture, values that emerge here include *love, kindness, patience, gratitude,* and *forgiveness.* They are the values that anchor some of our most familiar religious narratives. It is natural for us to think of them, therefore, as the byproducts of an advanced civilization. But these very same values characterize maternal relationships among all mammals. If they did not, there would be no way for their young to survive. In the calendar of evolutionary history, they precede all our human narratives—indeed, precede language itself. Our deepest core values, in other words, do not derive from distinctly human activity. They were already in place prior to our arrival on the scene. It is only subsequently that human cultures engaged with their presence to develop our distinctive, language-inflected strategies for living.

Just as importantly, these values manifest themselves today in the mammals we know and love. Our pets do not "appear" to love us because they have learned to mimic our moods to garner our affections. They truly do love us and forgive us because that is how they—and we—are wired from birth. We both went through the same nurturing experience, and we both carry that experience forward with us into adult life.

Given that the values we share with other mammals were not developed or held in place by language or narrative, what does hold them in place? The answer is emotional rewards delivered by the very same hormonal systems that underpin the operations of desire and fear. In the domain of mammalian nurture, there are at least three separate and distinct sets of feelings that provide positive reinforcement for nurture-based values. They are *pleasure, happiness,* and *joy.* Each plays a very different role in the subsequent development of human values, so it is worth taking time here to tease out the differences among them.

In the domain of nurture, we associate pleasure with nursing, feeding, cuddling, tickling, and the like. Moms and kids alike do this stuff in part just because it feels good. It is inherently sensual and closely linked to the fulfillment of desire and the hormonal reward system that enables it. All such systems operate on a presence-versus-absence basis, meaning they are either *on* or *off*. The off state for pleasure is not pain—that is a different kind of "on" state. Instead, it is simply the absence of pleasure, something closer to boredom. Like all off states, boredom is characterized by a restlessness as the organism seeks to rekindle the on state, and that restlessness motivates actions upon which social selection, in the form of either acceptance or rejection, can act. Teasing, for example, is behavior that tests these boundaries. Through it and other forms of testing, over time a subset of socially acceptable pleasures emerges in conjunction with a set of behavioral norms. From these come the foundational values upon which culture will build. All this is made a bit more complicated by sexual desire, but at the end of the day that too is an extension of the same forces socialized through the same mechanisms.

By way of contrast, *happiness* is distinct from pleasure because it is experienced socially, not sensually. Here hormonal rewards are linked to the welfare of the group. Whenever individuals signal to one another their state of well-being, this in turn reinforces a sense of well-being in everyone else. All our fear-based fight-or-flight mechanisms can take the day off. The experience of happiness that ensues creates a communal state, a sense of belonging that is most intensely felt in the context of immediate family but reaches outward as well to incorporate extended families, neighborhoods, tribes, teams, and the like. The strategic idea here is that happiness, unlike pleasure, gives everyone in the community a stake in everyone else's well-being. A single unhappy individual can turn the happiness state off for the entire group—giving rise to such proverbs as "You are only as happy as your unhappiest child."

And then there is *joy*, a third system of hormonal rewards that is neither sensual nor social. Here we have to tread carefully. Recovering lost joy is one of the dominant narratives of our secular age. One need only think of Thomas Wolfe in *Look Homeward, Angel* calling out "O lost, and by the wind grieved ghost, come back again!" He is expressing the collective nostalgia of the modern age for the joy and consolation of a departed religion. In a secular era, we still need that joy—perhaps more than ever—so we are going to expose it to a lot of pressure when we reach the top of our staircase model and seek a transition from metaphysics to ethics. It is imperative, therefore, to get as clear as we can about joy's mammalian roots prior to subjecting it to the impact of language, narrative, analytics, and theory.

To begin with, what makes us think that joy manifests itself prior to the emergence of human culture? The evidence for such joy comes from witnessing the phenomenon of *innocent play*. Whether it is a deer splashing around in a puddle, a school of spinner dolphins breaching over and over again alongside a boat, a cat with a ball of string, or a toddler enchanted with falling rain, we can—thanks to YouTube—see the same spontaneous, gratuitous delight play out over and over again.

We call these moments *joyful* because the intensity of the delight greatly exceeds the intensity of the stimulus. Indeed, the two are crazily out of proportion to each other. Hence, we get the sense of a gift, something wonderful that is being conferred upon us, something that we have not earned. That's what creates the arousal and the repeated behavior. It's what makes us laugh when we watch. We are hormonally wired to seek joy. The challenge we face as humans living in an age of digital distraction is that many of us have lost touch with joy's on/off switch—but more on that later.

In terms of locating joy within a domain, if it is neither sensual nor social, then what is it? The first word that comes to mind is *spiritual.* That word, however, entails a host of connotations from higher up on our staircase that we do not want to invoke at this point. Is there any non-anthropomorphic meaning to the word *spirit* when we restrict it to higher-order mammals? One point to highlight is that joy, specifically as it manifests itself in play, engages all aspects of an organism at once. The body is deeply involved, but so is consciousness, and so is emotion. There is an integration of response that is hard to tie to any one subsystem. In that context, perhaps we can use the word *spirit* to reference an integration of faculties, whether such a system persists in its own right or comes into being only at the moment of integration. In other words, at minimum we need to acknowledge a *spirit of play,* whether we locate it in the organism per se or in the phenomenon unfolding.

The most important point to take away from this is that joy is part of our mammalian heritage. Like pleasure and happiness, it comes from the domain of nurture, for that is where play is indulged. However, as a child matures to take on adult responsibility, there is much less room for play. Absent parental protection, natural selection now makes immediate and pressing demands on the individual, and the Darwinian mean must incorporate more of the fight-or-flight side of the desire-fear spectrum. There is a necessary loss of innocence, one that can lead to losing the ability to access joy. This challenge too is not specific to humans but rather is inherent in our mammalian heritage, as we can see when we turn to the second locus of value creation, *group governance.*

Mammalian social groups are typically governed by a hierarchy of power led by an alpha individual, usually male. While there are recurrent challenges to the leader's status, particularly during mating seasons, the hierarchy itself is remarkably stable. Why? Why do lesser enfranchised individuals in a group endorse and support a regime that allocates greater rewards to another?

Once again, it is because we inherit from our mammalian roots a series of hormonally reinforced systems of reward that engender a set of values that support maintaining the status quo, a kind of social homeostasis. These systems align with the lower levels of Abraham Maslow's hierarchy of needs: *security, community,* and *recognition.* Each delivers Darwinian benefits to mammalian groups such that it emerges naturally from the competition for scarce resources under adversarial conditions. The strategic principle is that mammals fare better as a group than as isolated performers, and that better-governed groups can outperform their peers.

Begin with security, the absence of which is exposure to risk. Risk not only increases the chances of death prior to propagation, it also causes organisms to consume metabolic resources at a fearsome rate. That is, the adrenaline-driven hormonal system that underpins the fight-or-flight response radically disrupts the body's natural homeostasis in order to reallocate resources for immediate use that are normally held in reserve. These reserves aren't infinite, however, and they are expensive to replace, so the less time your cat spends on a hot tin roof, the better. When group behavior evolves to reduce individual exposure to risk, hormonal systems can redirect our energies to more homeostatic ends. This is one of the reward systems that keep individuals committed to the group even when they have to subordinate their own desires to those of others.

Community, the sense of belonging to a valued group, builds on this foundation to reinforce self-sacrificing behavior. Self-sacrifice viewed superficially can appear non-Darwinian at the individual level, but it increases the probability of survival of the tribe. In this way, it helps perpetuate the gene pool, including the genes of the individual in question or of close cousins. There is an unresolved debate about whether animals truly demonstrate altruism, but this misses the point. Self-sacrifice broadly understood is a much more everyday occurrence than life-or-death decision-making. When individuals willingly accede to the status quo, they inevitably curtail their own desires in service to a social good. Whether this choice is made consciously or not, it is strategically beneficial to all, and natural selection will favor it. That is, reinforcing the stability of the hierarchy of governance helps maintain the status quo.

But that raises another question: Why would the stability of the status quo be favorable to evolution? What about innovation? How can species evolve if they cling to their current state? Here we need to distinguish between two kinds of change—*continuous* and *discontinuous*—driven by two kinds of innovation—*sustaining* and *disruptive*. Sustaining innovations that lead to continuous improvement represent the norm in evolutionary history. Species stay roughly the same but get better at what they do over time. Beavers build better dams, antelopes build better early warning detection systems, finches develop more fit-for-purpose beaks, and wolves learn how to hunt better in packs. As long as the environment is reasonably stable, this gradual progression is the winning path. Steady as she goes far outperforms taking a drastic leap.

Occasionally, however, the environment itself becomes radically disrupted. This may be due to a cataclysmic event (say, a major volcano or a meteor strike) or the tectonic reformation of a geological boundary that results in the introduction or elimination of one or more new species. Whatever the catalyst, the net effect is the same: traditional strategies for living no longer succeed in the new context. This creates a crisis that favors mutation. Most of these mutations fail, but at little ultimate cost since the owners of these genes were doomed already. But occasionally one succeeds, and from that emerges a new strategy for living embodied by a new species (or a meaningful variation on an old one). This new one is better adapted to cope with the altered environment. Wings, for example, may originally have developed to regulate body temperature by bringing circulating blood closer to the surface of the body (an early example of air cooling). Under conditions of stress, however, wings can flap their way into a new mode of locomotion—and voilà, we get flight. But that does not happen very often.

The interoperation of sustaining and disruptive innovation creates a pattern in evolutionary history called *punctuated equilibrium*. This consists of long periods characterized by a relatively stable set of species interrupted from time to time by a massive explosion of new species, such as the one that kicked off the Cambrian era. Such a pattern seems to call into question the validity of Darwinism itself, because Darwin did not have a good model for incorporating the impact of disruptive innovation. But today, thanks in part to our deeper understanding of how bacteria and viruses dramatically increase their rate of mutation when under threat of extinction, we now can expand the model of evolution to incorporate both rhythms of change.

Returning to the more homeostatic model of continuous improvement within a status quo, individuals competing within a social community increase the group's overall competitive competence. The competition stimulates the innovation, and the community preserves the group. The mechanism that drives the whole process is a hormonal reward system tied to social recognition of the leader, consisting of deference, preference in mates, priority in eating, solicitous grooming behavior, and the like. Such rewards are sufficient for alpha winners to forgo pursuing their competitive advantage to the limit, allowing others to have a share in the spoils. This reinforces the bonds of community, thereby ensuring more effective competitive responses when the community is under attack. It is how any good pecking order operates, even in Hollywood.

To sum up, the three values we have just reviewed—security, community, and recognition—form the foundation of the mammalian social contract. As with all things mammalian, we humans inherit this contract as a condition of our existence. To be sure, via the resources of language, narrative, analytics, and theory, we modify it in highly distinctive and differentiating ways. But the core of the contract comes from below. We all start with the same clay, regardless of how different the pottery we end up making may be.

Such differences first come to light with the emergence of culture.

Stair 7: Culture

The Transformation of Evolution: From Genes to Memes

The word *culture* in everyday speech connotes human achievement at its highest levels of sophistication. In this context, culture is deeply imbued with language, narrative, analytics, and theory—all stairs yet to come in our staircase model. So let us say that culture in our normal sense of the word represents something that is at the end of an evolutionary journey. Our purpose here is to determine how it got its start. What, in other words, does *culture* mean, if you strip away all of these higher elements, including language?

Let me propose the following comparison as a point of departure:

Just as a *genome* replicates a set of strategies for living that is *biologically maintained* and transmitted from generation to generation via *genes*,

So a *culture* replicates a set of strategies for living that is *socially maintained* and transmitted from generation to generation via *memes*.

Upon this analogy rests the transformation of evolution from the realm of genetics to the realm of ideas. It is the "missing link" that joins matter to mind.

In the context of culture, a *strategy for living* consists of a socially remembered pattern that is detected in a present situation, thereby activating a characteristic behavior that has some probability of achieving a desired result. These strategies may in part derive from instincts that are genetically transmitted, but they are also developed, learned, and refined socially, thereby allowing a local population to develop competitive advantages under conditions of natural selection.

Examples of culture understood in this sense can readily be observed in the tool-using behaviors of animals. Chimpanzees use twigs to fish for insects in a hollow log. Sea otters break open mollusks by crack-ing them against a rock balanced on their stomach. Gulls drop those same mollusks from high altitudes to crack them open on a rocky ter-rain below. In each case, whatever instincts are involved, the behavior is enhanced by observing and imitating the actions of others. Thus, a spe-cific tactic can emerge in one cohort of a given species and not emerge in a second one living nearby in essentially the same habitat. The culture of one group has superseded that of the other.

Culture is transmitted through the medium of memes. A *meme* is a unit of imitation that incorporates a unit of strategic behavior. The word derives from the Greek word for imitation, *mimesis*, which allows us to acknowledge Aristotle's seminal contribution to the idea. In the *Poetics* he points out how imitation is fundamental to how children play and then goes on to relate it to the esthetic pleasures of art. In so doing, however, he leapt from the alpha to the omega of culture. To get back to the alpha, to the origins of culture, we need to look more closely at what those children are doing.

Playing, as a social behavior, emerges in mammals but not in rep-tiles. This is not for lack of cognitive ability in the latter, but because only the former nurture their young. Nurture is invaluable to the transmis-sion of culture because it creates a safe space in which the young can innocently experiment with strategies for living. This is the evolution-ary function of play: over time, it results in learning. Subsequently, that learning can be supplemented by teaching, an extension of nurturing that goes beyond maternal care to incorporate participation by other members of the group. In every instance, strategies for living are being transmitted between generations through social activity. That is the evo-lutionary function of culture.

Culture can be defined, then, as the set of strategies for living that characterize any given community. The mechanism by which culture is transmitted is the meme, a unit of imitation that incorporates a unit of strategic value. Memes are directly analogous to genes in their ability to transmit strategies for living between generations.

This insight, as well as the coining of the word *meme* itself, we owe to Richard Dawkins. It is nothing less than a stroke of genius. By align-ing memes with genes, we can extend the principles of evolution beyond the material world of biology to incorporate the immaterial world of language, narrative, analytics, and theory. That will shape our entire approach to the remaining stairs in our staircase. Here's how the analogy plays out in full.

Darwinian forces act on memes exactly as they do on genes—through natural and sexual selection. Natural selection, you will recall, privileges strategies for living that succeed in the competition for scarce resources. This is survival of the fittest at work. It is half of the Darwinian formula, the goal of which is to live long enough to propagate. The other half is propagation itself, and that requires securing the cooperation of a mate. This is where sexual selection comes in, and it drives a different set of investments and behaviors, often at odds with the goals of natural selection, but critical to passing genes on to the next generation.

When we apply this same model to memes, we see that natural selection will select for strategies that work, that are successful. Sexual selection will select for strategies that are appealing to imitate, that are popular. Both are required to evolve a culture. If the strategy is wise but unappealing, if it teaches but does not delight, then it will struggle to gain an audience and fail to garner enough adoption to survive. Think of any number of instructional videos before the arrival of Sal Khan and the Khan Academy. On the other hand, if a meme delights but does so in folly, then it will spread like crazy but end up leading its practitioners to a bad end. Think of any number of "Don't try this at home" videos on YouTube. When, however, the two selection dynamics combine successfully, then memes mature and evolve like species of animals. Different ones take hold under different circumstances. Different communities embrace different memes. Each community develops its own culture. This holds as true for animals as it does for humans, albeit on a more limited scale.

Summing Up

With our discussion of culture as a mammalian artifact, we have completed our survey of the portion of the staircase that is prelinguistic. The emphasis throughout has been to accentuate our rich heritage as mammals and the legacy it bequeaths to us. We are born not only with desire and consciousness but also with a predisposition to values and culture. These come to us naturally because they were selected for naturally. We are as much artifacts of them as they are artifacts of us.

It is critical that we establish this foundation before we transition to a linguistic, human-specific domain. Language, narrative, analytics, and theory can lead us astray in many ways. They can make us believe, for example, that culture is a scarce asset reserved for a refined elite when in fact it emerges naturally from social life itself. They can make us believe that we need to derive our core values analytically from first principles when in fact they antedate our entire species. They can make us believe that consciousness isolates us from material reality when in fact it is rooted in intentions and desires that connect us to the physical world. They can make us think that intention and desire themselves are the enemies of reason when in fact they are its parents.

The Western intellectual tradition has a long history of such confusion. It became so enamored of the power of its higher faculties that it lost connection with its lower ones. The purpose of this chapter has been to reassert that connection. The purpose of the next is to use this foundation to realign the higher faculties accordingly.

CHAPTER 4

The Metaphysics of Memes

This is our third and final chapter tracing the staircase as it evolves in complexity from bottom to top. Here we take on the stairs that are specific to humanity itself: language, narrative, analytics, and theory.

The Infinite Staircase
(The Metaphyics of Memes)

11. Theory
10. Analytics
9. Narrative
8. Language
7. Culture
6. Values
5. Consciousness
4. Desire
3. Biology
2. Chemistry
1. Physics

Recall that the claim of this model is that each stairstep is a prerequisite for the emergence of all above it and that no stairstep is a prerequisite for the emergence of any below it. In this context, the first chapter focused on the three lowest stairs, exploring the *metaphysics of entropy* and concluding that entropy itself is the physical mechanism that powers the chain top to bottom. That is, the physical purpose of life on Earth is to help the planet shed the energy that the sun bombards upon it on a daily basis by increasing entropy in every way it can. The greater the complexity of a life form, the greater its ability to create entropy. By this metric, as we noted earlier, human beings outperform every other life form by far.

The metaphysics of entropy, however, does not by itself explain how life develops with increasing complexity—it only says that such a turn of events will be supported if and when it appears. To understand how complex life forms actually do develop, we turned to the *metaphysics of Darwinism*. Here we traced the emergence of complexity from desire to consciousness to values to culture, the common thread being an evolution in the ability of populations to create and execute *strategies for living*. The more complex these strategies are, the more adaptable the species. By the end of this section of the staircase, we were focused exclusively on mammals because they are able to transmit strategies for living across generations socially, not just genetically. We adopted the term *memes* to represent units of strategy so developed and so transmitted, and we situated their emergence prior to the emergence of language.

Now it is time to revisit the concept of memes in the context of language. Language is a massively disruptive innovation, and we are going to trace its impact, first in its own right, then as an enabler of all the stairs above it. The common thread throughout this journey is the evolution of increasingly effective, socially communicated strategies for living. Thus, we shall call this final leg of the journey the *metaphysics of memes*.

And the punch line to our story will be a simple one: As human beings we come into this world a product of our genes; we go out a product of our memes.

What Are Memes?

While we have already discussed this concept in some detail, we did so in a prelinguistic context, where we adapted it to an animal theory of mind. Now, as we transition to the domain of language, we can represent memes as we humans experience them. We call them *ideas*.

As ideas, memes have two defining attributes. First, they have strategic value, meaning they help us size up a situation or enable us to take some useful action within it. Second, they are attractive, meaning they engage and enlist us to adopt their point of view and to share it with others. We try on the idea for size to see how it fits. The more we like the fit, the more we promulgate the meme, and the more rapidly and broadly it disseminates through our community.

Successful memes are contagious. They recruit us into their service, inspiring us to commit to certain causes and act in distinctive ways. Religious beliefs are memes. So are political platforms. So are family habits. Memes are the building blocks of social identity, the stuff that defines who we are (or at least who we would like to be). But they are more than that as well. All science consists of memes, all history, all literature, all social science—basically all knowledge. It's all memes. It all consists of ideas that have the same two defining attributes: strategic value and viral attraction.

Now, to be fair, I never caught the calculus virus, just as others never caught the medieval literature virus or the golf bug. But everyone catches some virus. Everyone is infected with some set of memes. These memes shape our behavior, our choice of social groups, and our sense of ourselves. From this perspective there is nothing more important than memes—they transcend even life itself. People die for memes, willingly.

So it is no accident that the metaphysics of memes underpins the top end of our staircase. This is as far as humankind knows how to reach. Our goal in these final four chapters is to get better clarity on just how far that is.

Stair 8: Language

The Fabric and Fabricator of Memes

Language is arguably the most disruptive innovation in all of evolution. Without it we are necessarily confined within our own space and time. With language, we are able to span continents and centuries, connecting ourselves to a much larger network of endeavor, participation, and meaning.

Language emerges as a mechanism for creating and communicating memes. Memes are culturally mediated artifacts of natural and sexual selection. How best to hunt, gather, cook, carve, court, or kill—evolution selects for them all. Language did not initiate the production and transmission of memes. That, as we saw in the prior chapter, emerged prior to language, among mammals and birds imitating actions that they witnessed or that were modeled for them. Language's contribution was to refine, accelerate, and amplify this process.

Children learn language with breathtaking ease and speed. Linguists have claimed this is due to a universal grammar that is genetically wired into the human brain as part of a language faculty. But how did the faculty itself emerge? The concept of memes provides the answer. Children learn language easily because, like all conscious beings, their brains are wired to learn strategies for living. Memes are strategies for living. The language faculty is organized around the requirements for communicating memes. It represents a fast track to accomplishing our desires. That's what makes language so easy to learn.

This is true of all six thousand languages currently extant on the planet. Despite their extraordinary differences in phonology, grammar, structure, and vocabulary, every single one of these languages is organized around performing five core meme-related functions. As speakers of English, here is how we experience them:

1. *Naming.* This is the role of nouns and noun-like phrases and clauses. We have been taught in school that nouns refer to people,

places, or things. But why do we want to refer to them? The theory of memes argues it is because they represent *bodies of forces that have strategic implications.* They can be objects of desire, agents to be wary of, or instruments to achieve our ends. Feeling thirsty? It would be good then to know nouns like *water, glass,* and *faucet.* Going to a friend's house? Best to know *car, road, gasoline,* and *GPS* along with some more abstract nouns like *address, time,* and *occasion.* Every noun we know earns its place in our vocabulary by naming a body of forces that matter to us in some way. In one context or another, it is strategically significant. If it were not, we would have no reason to remember it.

2. *Describing* and *Prescribing.* This is the role of verbs and verb-like phrases. We use them to refer to *states of being* and *actions.* Both are attributable to the impact of some body of forces. That is what makes for the fundamental connection between nouns and verbs. What good is it to know the word *glass* unless you also know it *holds* water in a way that makes it easy for you to *lift* it and *drink* it? States of being, in general, prompt actions (the glass is full, and you are thirsty). Actions, in general, change states of being (the glass is empty, and your thirst has been quenched). Wherever behavior is strategic, language aligns with it directly.

3. *Modifying.* This is the role of adjectives, adverbs, prepositional phrases, and various clauses that attach themselves to noun or verb phrases. We use modifiers to contextualize, characterize, specify, or amplify *our understanding* of a body of forces, a state of being, or an action that involves one or more bodies of forces changing one or more states of being. Is the water cold? Is the glass on the table? Are you drinking the water slowly? Do you feel refreshed? The more accurately we modify our linguistic representation of things, the more precise our strategies for living can become.

4. *Connecting.* This is the role of conjunctions and conjunctive adverbs and phrases. We use them to describe or propose *relationships* among entities referenced by the three classes of language noted above. Thus, *when* you got thirsty, you filled *and* drank a glass of water, *but* you forgot to wash it *and* put it away.

Conjunctions let us piece together the complex relationship maps that underlie advanced strategic behaviors.

5. *Claiming*. This is the role of sentences and other forms of predication that *assess, acknowledge,* or *advocate* a strategic understanding of a situation or a behavior. *You look thirsty. There is a glass in the cupboard. You should get yourself something to drink. I think there is OJ in the fridge.* We use claiming to propose a point of view as a precursor to undertaking actions, as well as to provide an analysis after the fact. This ability sets human culture far apart from that of any other creature.

These five capabilities working in tandem allow humans to create, communicate, execute, and critique strategies for living. They are what make us effective in the world. At their core is the ability to describe a situation and then propose, prescribe, or actually make a response. All our essential claims are verb-based predications made with respect to noun-based named entities. Modifiers focus and tune these descriptions and prescriptions, and conjunctions allow us to build increasingly complex scenarios out of simpler component elements. That, in essence, is how language equips us for living.

Thus far we have been describing the *semantic* dimension of language. In this domain, language is instrumental, logical, and pragmatic. Its performance criteria are based on the outcomes of the strategic behavior it has conceived. The semantic dimension of language is universal and equates to Noam Chomsky's "hard wiring" of linguistic competence in the human brain. That is, all languages perform these semantic functions, and all humans are predisposed to learn language because language aligns directly with strategies for living.

To call this property of language a universal grammar, however, is misleading. Language requires a much bigger framework to capture its operations across all its various dimensions. Specifically, a comprehensive treatment of language must address at least six distinctly different domains of competence:

- Semantic: *underlying meaning*
- Grammatical: *surface structure*

- Performative: *speech and writing*
- Expressive: *self-expression*
- Rhetorical: *interpersonal engagement*
- Imaginative: *speculative exploration*

Each of these domains is governed by a distinctive discipline specific to it, and all contribute to language's dramatically disruptive impact on human evolution. We have already addressed the domain of semantic competence. Now let's look briefly at the other five.

Grammatical competence focuses directly on the structure of language itself. It is by nature conventional. Each language exhibits a unique set of rules governing construction of its various forms. Grammar in this sense is the opposite of universal—it is inherently local and idiosyncratic. Nonetheless, in every language, grammar does serve two sustaining purposes. First, it maintains the integrity of the language as a social medium of expression. It ensures consistency across users and use cases such that the language can be learned and mastered by an entire community. Second, via minor or even minute variations in grammatical forms, grammatical competence signals the social class and ethnic background of the speaker or writer, both of which provide key inputs to the performative and rhetorical competencies discussed below.

Because grammar is inherently idiosyncratic, it is a complex and complicated field, and we will do well not to wander too far into it. Why, for example, are two grammatically similar sentences like "John is easy to please" and "John is eager to please" so semantically different, and how do we know how to tell the difference between them? This is a realm for specialists interested in the subject of grammar per se and, for our purposes, has little bearing on the metaphysics of memes. To paraphrase Obi-Wan Kenobi in the first Star Wars movie, these are not the droids we are looking for.

This is largely true as well of the next domain of linguistic competence: *performance.* Performative competence represents the physical ability to produce and share language, something that is specific to each individual. Physiologically, the human palate and our other organs of vocalization set us apart from higher apes in our ability to articulate a nuanced variety of vowels and consonants. Similarly, the human hand's fine motor capabilities make writing, signing, and keyboarding feasible at scale. Both are universal enabling mechanisms that have shaped the evolution of our linguistic competence overall. At the same time, for any particular language, the sounds, signs, and graphs that constitute its performance are arbitrary and conventional, combining with grammar to create significant switching costs for any language community. These switching costs may be an artifact of our tribal heritage—indeed they may well have contributed to it by reinforcing communal bonds and dependencies. That said, the impact of performance competence on the metaphysics of memes is also negligible.

Expressive competence, by contrast, is crucial to the development and maturation of human personalities. It manifests itself very early in life through inarticulate but highly emotional coos and cries. As months pass, coos and cries morph into words, phrases, and sentences (some of which are in make-believe languages), and eventually mature into whole trains of thought. The goal of these trains is to express and explore one's current state of being and to share it with others. Thus, language represents mood, desires, thoughts, and fears, be it in daydreaming, soliloquies, journals, poems, songs, or conversations. It is a medium for self-discovery, a mirror in which to see oneself, as well as a bridge to connect emotionally with others. For social animals, it represents a kind of ground zero.

Rhetorical competence extends the bounds of expression from self to others. In so doing, it competes with semantic competence as to which should be acknowledged as the more empowering, and it has done so ever since Plato took issue with the Sophists. Whereas the primary focus of semantics is on *power with respect to acting in the world*, the primary focus of rhetoric is on *power with respect to engaging with others*. If semantics describe and prescribe our causes, rhetoric helps us enlist others to support them. Semantics is tested and refined through a kind of natural selection based on the success of our actions. Rhetoric is tested and refined through a kind of sexual selection—namely, social selection—based on our success in winning others over to our points of view.

The domain of rhetoric ranges from sound to sense. At the level of sound—rhyme, rhythm, alliteration, and the like—rhetoric makes memes more memorable. That is why it is easy to recall that *a stitch in time saves nine* or that *early to bed, early to rise, makes a man healthy, wealthy, and wise.* But sound can go to the emotional heart of things as well, be it joy—"i thank You God for most this amazing day"—or ennui—"In the room the women come and go / Talking of Michelangelo." Rhetoric here is engaging an audience through empathy and shared vulnerability.

Taking the craft of engagement further, while semantics focuses us on the *denotations* of words, rhetoric attends to their *connotations.* Connotations are webs of associations that position, contextualize, and value the same basic concept in varying ways, as in "I am *brave*, you are *daring*, he is *reckless.*" In addition, rhetoric can vary the tone of a message by incorporating honorifics or using privileged grammatical forms. In this way, language users are better able to negotiate social relationships of power as they seek to enlist others in their endeavors. When asking for funding from a venture capitalist, for example, successful entrepreneurs don't normally say, "Why doncha, like, gimme a million bucks, dude?"

Finally, there is another dimension of rhetoric that we call "figures of speech." In the Renaissance period, these figures were largely understood to be decorative in nature. They were ornaments to enhance a speaker's or author's efforts to engage and enlist: "Shall I compare thee to a summer's day? Thou art more lovely and more temperate." Subsequently, however, both Romantic and modern sensibility took a deeper dive here, especially in reference to metaphor, and that, in turn, gave prominence to a sixth discipline of linguistic competence: imagination.

Imaginative competence begins with perceiving likeness. These likenesses may initially be superficial—"Do you see yonder cloud that is almost in the shape of a camel?"—but over time the perception of likenesses goes increasingly deeper:

- A car is like a horseless carriage.
- A pimple is like a tiny volcano.
- A company is like a football team.
- DNA is like a computer program.

Such likenesses can become the basis for entire strategies via the notion of a root metaphor or an extended analogy. This book, for example, is structured around an analogy to a staircase, something we will come back to in our closing section about theory. M. H. Abrams wrote a masterful work of literary criticism, *The Mirror and the Lamp*, that showed how neoclassical culture saw the role of art as holding up a mirror, whereas Romantic culture saw it as holding up a lamp. Plato's metaphor of the cave, Descartes's metaphor of the mind in the machine, Nietzsche's metaphor of Apollo versus Dionysus—all drive acts of imagination that transform complication and complexity into clarity and coherence.

Over time, metaphor seeps into the very fabric of language itself. We say a marketing campaign has gone *viral*, we trace our family *trees*, we *fish* for compliments, and we *browse* the Internet. The cumulative impact of such associations is to draw attention to patterns in one realm based on having witnessed them in another, thereby implying that the strategies that create success in the known realm might well be applied to the unknown one.

Metaphoric language, in this context, is hugely disruptive. Its impact on the universe of memes is analogous to the impact of mutation and recombination on the universe of genes. That is, it produces variations against which natural selection (semantic) and sexual selection (rhetorical) can operate. Google Search is like an oracle. Uber is like a transportation genie. Airbnb is like knowing friends everywhere. To be sure, the overwhelming majority of new memes created in this way do not survive, but a few proliferate, and some become progenitors of whole new species (worldviews). It all begins with an act of imagination.

To sum up, of the six domains that make up language as a whole, three in particular separate humanity dramatically from our closest evolutionary relatives:

1. *Semantic competence* enables us to represent, prescribe, enact, and critique strategies for living with far greater precision than any other creatures. *"I am hungry, and I see two antelopes over by those trees."*
2. *Rhetorical competence* enables us to engage more deeply and enlist more broadly a diverse population of others to support our strategic intentions. *"Wouldn't it be great if we had some food for our families tonight?"*

3. *Imaginative competence* enables us to conceive, invent, plan, and forecast strategic actions and outcomes that are not manifest in our present situations. *"If you sneak up on the right, and I on the left, we can corner them by those rocks over there."*

As the hunting tactics of wolves make clear, all three of these capabilities were already latent in mammalian consciousness and culture, but without language they can find only the most primitive expression. Language changes the game. It enables strategies for living to transcend time and space, thereby increasing both their scope of application and their likelihood of persistence. We are the creatures of those strategies.

In the remaining three sections of this chapter, we will explore three distinct domains of language usage: narrative, analytics, and theory. Our goal will be to show how, taken together, they underpin and make possible the whole of human culture.

Stair 9: Narrative

Inventing Strategies for Living

Language, left to its own devices, is tactical by nature, an instrument for everyday living. It helps us negotiate our tasks and relationships day to day. It is only with the emergence of *narrative* that it becomes truly strategic.

Narrative is synonymous with storytelling. Stories are our fundamental means of making sense of the world, our first pass at understanding cause and effect. Via stories we describe a sequence of events, and through analytics we impose an interpretation upon those events—the moral of the story, if you will. These interpretations have the potential to apply to a whole class of events, to become proverbial. Aesop's fables provide a simple example of this effect, Dante's *Divine Comedy* a complex one. Both authors, however, are advocating for the strategies for living exemplified in their narratives.

Literature enables us to entertain strategies for living in a hypothetical space where they can be explored provisionally. Could we act like Huckleberry Finn? Holden Caulfield? Elizabeth Bennet? Hester Prynne? Harry Potter? More importantly, could we learn from them? For embedded in their stories are strategic principles that can be extracted through analysis. That is why narrative precedes analytics in the staircase model—it gives analytics something to analyze.

As with language, there are a number of component elements that make up narratives, and here too each one performs a critical function in representing a strategy for living:

1. **Character.** Characters are *the agents* who embody and enact strategies for living. When you identify with characters, you are imaginatively participating in their strategies, experiencing their risks, and sharing in their rewards. This gives stories enormous motivational impact.

2. **Setting.** Settings *frame the context* within which a given set of strategies will be tested. They function like a game board on which pieces can be moved, providing both resources and obstacles for characters to engage with. Also like a game board, they imply a set of rules that govern the universe of possible choices and outcomes. Narratives that keep faith with their implied rules are judged to be "realistic" even when the rules themselves are contrary to the normal circumstances of everyday life, as tales of fantasy and science fiction bear witness.

3. **Incident.** Incidents allow characters to *demonstrate their strategies* through tactical moves, the outcomes of which measure success or failure in that particular situation. As narratives progress, the challenges become more difficult, and the strategies are put under increasing pressure to evolve.

4. **Plot.** Plots pass *overall judgments* on strategies for living, primarily in terms of final outcomes. The plot is "done" when the ultimate challenge has been met, regardless of the outcome.

5. **Dialogue.** Dialogue reveals characters' *motives and intentions* as well as their understanding of their situation, both as they use speech acts to accomplish their ends and as they provide, intentionally or ironically, added perspective on their strategies for living.

6. **Narrator.** Narrators *modify our perception* of the previous five elements in order to develop more nuanced representations of the strategies being exemplified. Their commentary represents a first step toward analytics.

7. **Style.** Style *engages and enlists audiences* through a wide range of rhetorical techniques that further shape our experience of the narrative.

Of these seven elements that make up storytelling, *plot* is the one that holds sway over the other six. We always want to know, *How does it end?* Endings confer summary judgments on the strategies for living of the characters we have been following.

Plots can be organized around four archetypes, described by Northrop Frye in *Anatomy of Criticism* as follows:

- *Comedy*, in which characters succeed despite their often inept strategies for living, in large part because forces outside their control intercede on their behalf. Think Ferris Bueller or *A Midsummer Night's Dream*.
- *Romance* (used in the literary sense of heroic adventure), in which characters succeed primarily based on the true merit of their strategy for living. Think Captain America or *Sir Gawain and the Green Knight*.
- *Tragedy*, in which characters fail, primarily due to a debilitating contradiction in their strategy for living. Think Captain Ahab or *Macbeth*.
- *Irony and Satire*, in which characters fail in large part because forces outside their control are set against them such that no strategy for living could succeed. Think Charlie Brown or *Waiting for Godot*.

These four genres have dominated storytelling from the very beginning, across all cultures and in all eras. Storytellers typically signal early on which genre they are following, and that evokes from readers and audiences a set of expectations that help create an engaging experience with a satisfying sense of closure.

The key point here is that genre determines outcomes, not the other way around. That is, what appear to be *descriptive* acts of language are actually implicitly *prescriptive*. Once you have entered into any given genre via your imagination, the range of actions characters can take is significantly curtailed, channeled into one of these four pathways, to be carried forward by—what? Some might want to say fate, or life, or chance, but the reality is that they are carried forward by allegiance to the conventions of the genre itself. *Genre* is the underlying force that shapes narrated events. It is the way imagination imposes order on a future that is forever indeterminate.

The prefabricated outcomes built into these genres may seem innocuous enough as long as they remain confined to fictional narratives, but in fact they bleed out into all our nonfictional ones as well. Their prevalence, indeed their irresistibility, is a consequence of sexual selection among memes. The four archetypes represent four kinds of stories we like to hear and love to repeat. And every time we repeat them, we tend to smooth out the narrative to make it a bit more compliant with the underlying archetype until in the end, absent critical analytics, all we get is repeated stereotypes. Welcome to the worlds of superheroes, horror films, and polarized politics.

To properly appreciate how pervasively genres permeate our world, consider the following classes of narratives that make up the very fabric of our everyday lives:

- **Religious.** Every religion is communicated via narrative. These narratives both describe supreme forces that govern the world and prescribe behaviors by which we can align ourselves with them. Religious narratives are all comedies, meaning that in the end things turn out for the best due to forces well beyond human control. Believers privilege these narratives by conferring on them a status of absolute truth, thereby making their ethical prescriptions unquestionable. Unbelievers, by contrast, experience these same narratives as a form of literature. They often see them as beautiful and profound, even holy, but not as absolute truth. A subset of unbelievers, usually self-identifying as atheists, can actually be infuriated by them, for they experience them as denying truths fundamental to their own core narratives grounded in contemporary science. What gets lost in all the subsequent arguing back and forth is that human beings need a "grand narrative" of some sort—an uber-story that contains all others within it—so that we can find our place in the world. That's what metaphysics is all about.

- **Societal.** One would normally use the word *social* here, but since all types of narrative are inherently social, we can use *societal* to refer to those narratives that create social cohesion. They can be of any of the four archetypes depending on current circumstances and the mood of the narrator. That is, a given community can perceive its fate as comic, heroic, tragic, or ironic. Regardless, their stories celebrate the differentiation of the group, creating a shared identity in which the entire community participates. In nation states, such narratives merge with history to create archetypes that shape national character and political themes. At every level, societal narratives further reinforce the community's cohesiveness by telling stories about "the others"—the people whom we are "not like." Under stress these are the stories that enable our worst atrocities—persecution, slavery, war, and genocide. The key point here is that they are *stories*, nothing more, nothing less.

- **Personal.** Personal narratives help resolve what for analytical philosophy can be a thicket of tough problems: What is identity? What is the self? What ensures the continuity of either or both? Narrative's answer to these questions is that the self is a character in a story, whose identity is defined by the strategies for living he or she embraces and exemplifies, and whose continuity is maintained by the narrator and the audience. In real life the narrator at the outset is typically a parent, but over time, narration is incorporated into the self, and we become self-narrating. Hence the serendipitous elegance of the question, *What's your story?* As elsewhere, personal narratives can form around any of the four genres. From this vantage personal narratives get exposed to a kind of literary criticism, both internally and socially, that tests both one's story (strategy) and one's behavior (tactics) against the expectations set by the archetypal genre selected. The result is a built-in homeostatic governance system that makes for a remarkably stable and adaptive outcome— namely, you. It is extraordinarily challenging for anyone to resituate their identity in a different genre. Nonetheless, it can be done, and when it does come to pass, it is usually described as a "life-changing" event, although "self-changing" would be more accurate. Think of Paul's experience on the road to Damascus, Michael Corleone in the Godfather series, or Ebenezer Scrooge in *A Christmas Carol*.
- **Business.** Narratives in business are fundamental to leadership, investment, and management. Leaders frame narratives around missions that engage and enlist employees, partners, and customers in a set of opportunities to address or problems to solve. To fund these ventures, they reach out to investors with another set of narratives about the financial gains from executing a particular strategy, causing the investors in turn to apply a stringent form of literary criticism of their own, probing for flaws that would cause these stories, and hence their investments, to unravel. Once investment is secured, executives develop a more tactical set of planning

narratives to allocate resources to align with achieving the vision embodied in the plan. Managers charged with executing these plans report back with narratives that describe actual outcomes in light of intended objectives, the goal in every case being to either win or learn. To be sure, there will always be numbers to crunch as well, but the meaning of those numbers must be supplied by an accompanying narrative.

- **Journalistic.** Journalistic narratives inform their audiences of events relevant to their interests and values. These are open-ended stories designed to be serially modified as additional facts emerge. In modern Western democracy, they are critical to ensuring informed political representation. As such, they are expected to be disinterested and objective, meaning the narrator makes explicit the sources of any value judgments made and refrains from covertly imposing a personal or ideological point of view. This is inherently challenging not only for personal and social reasons but also because the archetypal genres underlying all narratives call for a kind of closure often beyond what the known facts warrant. The best journalism embraces this challenge and pursues its ends as transparently as possible. Much conventional journalism falls considerably short, capitulating to the demands of archetype in order to secure popularity, thereby winning the war of sexual selection by defaulting on its commitment to natural selection. Who cares if the news is fake if people read it and advertisers pay for it?

- **Historical.** History seeks to discern patterns of cause and effect in the course of human events, which it then communicates via narratives. The narrator is the historian, and the historical agents are the characters. The "great man" theory of history (including the sexist moniker) continues to thrive, with its narratives merging into biographies that are frequently shaped by genre ("rags to riches" being a comic or heroic plot, "fall from grace" a tragic or ironic one). By contrast, impersonally grounded histories founded on archeological evidence, economic data, or actuarial statistics focus more on setting and incident, often celebrating the anonymous lives of less privileged social classes. In so doing, however, they forgo some amount of "reader appeal" by sidestepping archetypal narratives. That said, all histories are narratives of one sort or another. Narration is the foundation of the discipline itself.

- **Scientific.** Science normally represents itself as an analytic discipline. In this context its primary use of narrative is to describe and record experiments. Here the scientist takes scrupulous pains to eliminate any source of personal bias, be that in the conduct of the experiment, in the collection, collation, and interpretation of the data generated, or in the presentation of the recorded results. An ideal scientific narrative in this context is perfectly transparent and genre free. But once these experiments have been properly registered and reviewed, then the more familiar form of cause-and-effect narrative can—indeed must—make its appearance. Narrative is essential to understanding and communicating cause-and-effect relationships, the how and why of things. Whether it is the Big Bang or the molecular biology of the cell, the weird behavior described by quantum mechanics, or the present danger of climate change, there is always a story to tell. Storytelling is fundamental to science, both to anchor conceptual understanding and to stimulate social action.

As these examples show, narrative, in one form or another, underlies everything we think and do. Its protean nature allows it to present and explore strategic dimensions across the entire spectrum of human experience. That's what earns it its own step on the staircase. It is a virtual laboratory where we may explore potential cause-and-effect relationships and select among them based on plausibility and fit with circumstance. In so doing we detect and impose patterns that shape our ever-unfolding, never predictable experience. Narrative is the stuff from which our strategies for living are made.

For those strategies to prove efficacious, however, they need to be tested and refined. That task takes us to the next stair up in our staircase: analytics.

Stair 10: Analytics

Testing and Refining Strategies for Living

The relationship of analytics to narrative is nicely captured in Daniel Kahneman's book, *Thinking, Fast and Slow*, where he explores the interplay between two modalities of problem solving—what he calls System One and System Two. System One, like narrative, exploits intuition to leap to rapid conclusions, whereas System Two, like analytics, follows a rigorous system of procedures to arrive at more reliable answers. Both have their place in a world that rewards snap decisions in some situations and considered deliberation in others.

Narrative exploits intuition to provide compelling explanations of cause-and-effect, ones that are readily grasped, easy to remember, and straightforward to act upon. These are the essential qualities of viral memes, and it is no accident that such narratives spread rapidly and gain adherents. But are these memes truly exemplary anecdotes or just urban legends? In other words, just what can we expect from a gluten-free diet, and how would we be able to know?

Enter analytics. These are disciplined procedures to discover, detect, define, describe, and measure the bodies of force you are dealing with. Regardless of what you are trying to do—build a boat, clean your house, understand your spouse, or improve your tennis swing—there is likely to be available information that can help you figure out what is happening, how it is happening, and why it is happening. Analysis is the process of bringing such information to bear on the situation at hand in order to validate or discredit the approach underway. It represents reason providing a check against imagination, desire, will, and, most of all, narrative.

But this is only the beginning. Over time analytical insights accumulate and aggregate into bodies of knowledge that become the subject of academic study. Whole departments form around the analytics of physics, chemistry, and biology; of psychology, sociology, and anthropology; of literature, history, and philosophy; of finance, medicine, and law. Scholars specialize in these disciplines. They earn and confer advanced degrees on one another. This endeavor quickly leads to analytics being applied to analytics, and eventually to analytics being applied in turn to those analytics, until the air becomes too rarefied for any but the sturdiest Sherpa to breathe.

That said, all told, the impact of analytics on humanity's well-being is nothing short of spectacular. Despite all the suffering and outrage plaguing contemporary life, conditions for human beings are vastly superior to what they were even a few centuries ago. This is primarily due to progress in analytical knowledge. Our narratives have not changed appreciably over the past two millennia, but our stock of useful information surely has.

We tend to give most of the credit for this to science, but to be more precise, it ought to go to mathematics. Without mathematics, science is just another exercise in natural language, with all the inherent ambiguities and imprecision such efforts are heir to. Mathematics, however, introduces a categorically different kind of language, one that is exquisitely precise and wholly unambiguous. Mathematical sentences interlock with a precision a Lego master can only envy. It is a far cry from the patchwork mosaics pieced together by natural language, where subjects are glued to predicates, propped up with modifiers, and mortared over with connective phrases. Claims anchored in mathematical analytics can build on each other with confidence. They can create edifices of extraordinary scale, exchanging insights and patterns across disciplinary boundaries with no semantic leakage.

Thus, Newton could build on Kepler, Maxwell could build on Newton, and Einstein could build on Maxwell. The frames of reference shifted, but the math remained solid. Contrast this with scholarship in any natural language discipline, be it in the humanities or social sciences. Great ideas in these fields do not build on one another. Rather, they emerge from dialectical interactions among opposing views, a never-ending dance of thesis-antithesis-synthesis, with each synthesis becoming a new thesis. Nothing is final; every connection is contingent. That, as my father liked to say, is no way to run a railroad. Unfortunately, for much of where we want and need to go, it is the only mode of transportation we have.

So how should one proceed? First, wherever mathematics can be applied, it should be applied. That is, we should continually challenge the boundaries of natural language with mathematical incursions. Two areas, in particular, warrant contemporary attention. The first is probability and statistics, very much along the lines Kahneman advocates. This discipline provides a corrective balance for our species' excessive confidence in the dictates of narrative. Citizens in a democracy have much more to learn from statistics than from the trigonometry and calculus that routinely round out high school math curricula. Han Solo may never want to be told the odds, but the rest of us need to know them.

The second area is machine learning, based on applying analytical algorithms to vast quantities of digitally recorded event data. This data comes from computer logs of every imaginable kind, be that clicking on websites, making mobile phone calls, capturing heartbeats, registering seismic signals, or sensing machine vibrations. You name it, if some sensor is detecting it, some log file is recording it, and some program can be written to read that log file. Of course, all that data is just a massive accumulation of haystacks unless you can impose upon it some scalable, signal-sensitive, needle-finding protocol. Fortunately, that is precisely what machine learning can do. It is still the early days in the development of this discipline, but already it has reengineered our approaches to genomic medicine, anti-terrorism, high-frequency stock trading, retail promotion, next-best-action recommendations, autonomous driving, preventative maintenance, content curation, and fraud detection—all areas where only a few years ago one would have argued that human intelligence was far better suited to the tasks at hand.

As powerful as it may be, however, mathematics is not a language that most of us can speak or read. Natural language is the medium in which we develop and critique our strategies for living. We are immersed in a world of memes competing for our attention and allegiance. The landscape is littered with fake news, fraudulent claims, and subversive propaganda. We must be able to sort through this trash to find useful knowledge. The more analytical intelligence we can bring to bear on this activity, the higher we can make the quality of our lives and livelihoods.

So what exactly does *natural language analytics* consist of? The foundation upon which the entire discipline is built is the *claim*. Every declarative sentence makes a claim. The subject of the sentence identifies the topic, and the predicate applies the claim to that topic: John is tall, cheeseburgers are fattening, and the Nile is a river in Egypt—all claims. The claims we are most interested in are those that imply an opportunity or a threat, an obligation or an entitlement, a pending pleasure or an impending pain. Such claims call for some kind of strategic response, and it is the role of natural language analytics to help us assess these situations, evaluate candidate responses, and choose the best from among them.

So how do we evaluate a claim? Assuming the claim itself is clear to us, we should inquire about *evidence*. What evidence is there that warrants belief in this claim? It turns out that whatever evidence is asserted, it too comes in the form of a claim. Cheeseburgers are fattening. Why do you believe that? Well, according to fastfoodcalories.com, a Whopper from Burger King has 650 calories. Ah, a skeptical person might interject, but why do you believe that claim?

It does not take long to realize that absolute skepticism leads to infinite regress, so how could one proceed in a more disciplined way? To begin with, evidence consists of claims we are willing to grant without further discussion. It is a pragmatic choice as to where one draws this line. There is no absolute boundary, only the ticking clock of mortality to remind us that life is a game played within time limits, something the ultimate skeptic chooses to ignore.

So suppose we take my fastfoodcalories.com data as valid evidence. Still there is the question of whether this evidence *warrants* the claim that cheeseburgers are fattening. Warrants, it turns out, are also claims. They claim there is a direct connection between the evidence cited and the claim being made. This relationship can typically be described via one or more intermediate claims along the following lines:

- The average number of calories a sedentary adult male should consume per day is between 2,000 and 2,600.
- The average turkey sandwich has about 300 calories.
- Substituting a Whopper with cheese for a turkey sandwich will add around 350 calories to your daily intake.

- Unless you do something to counterbalance this, those calories will turn into fat.

Testing the interactions among claims, evidence, and warrants represents the core of natural language analytics. It was usefully dissected by Stephen Toulmin in his 1958 book, *The Uses of Argument* which presents the model for analytics upon which I have been drawing thus far. There are further subtleties to this model, as he and others have described. Take the word *unless* we inserted in the fourth warrant just above. That represents a *qualification*, which modifies a claim to increase its precision and accuracy. One could also make a *concession* to a potential counterclaim that not everyone would get fat, perhaps along the lines of "Athletes who exercise strenuously, of course, will burn off these extra calories long before they convert to fat." Conversely, one could also provide a *rebuttal* to a less credible counterclaim along the lines of "One might argue that 650 calories are not all that much, but when you throw in the fries and the Coke, your lunch weighs in at over 1,000 calories, and you are well on your way to a much bigger belly."

Such are the devices of argumentation, and mastering them is part of the blocking and tackling of natural language analytics. They are built upon a fundamental link between claims and evidence, assertions and data, interpretation and fact. To be useful and effective, argumentation must be grounded in a social contract that respects these connections. When politicians deflect all criticism as fake news, or when their apologists refer to alternative facts, they are breaking this social contract, one that has been fundamental to Western democracy as it has been practiced for the past two centuries. Analytics cannot perform their function under such corrupt regimes, so these regimes must be repudiated in full, regardless of how much effort that might entail.

In addition to argumentation, there is another dimension to analytics we need to examine. This one experiments with new relationships as opposed to verifying established ones. Such experimentation is prevalent throughout the mathematical sciences. Time and again, extrapolations from purely mathematical relationships have predicted, and subsequently led to, the discovery of novel phenomena, be that in physics with the Higgs boson, or cosmology with Einstein's prediction that light rays would bend in a gravitational field. With natural language analytics, there is an analogous capability to do the same. Here the heavy lifting is done by similes, analogies, and other forms of linguistic comparison, what we will call collectively *metaphors.*

Metaphors are one of humankind's most powerful problem-solving tools. They juxtapose two situations, one of which is mysterious, the other well known. A metaphor makes the claim that the structures of the two situations are sufficiently similar that one can apply the strategies learned from the familiar situation to address the challenges posed by the unfamiliar one. Sometimes this is communicated in an offhand or oblique way, as when one would talk about "sexual magnetism" or the "slippery slope" of moral compromises. But in other cases, the metaphor is both explicit and overt.

Consider the following two metaphors:

1. "Thoughts come to mind like bubbles rising from the bottom of a pond."
2. "Thoughts flow in a never-ending stream."

Both address the part of thinking that is involuntary, a rich source of ideation that is more witnessed than performed. Their strategic implications for how best to leverage that source, however, are different. If thoughts are like bubbles, then sitting quietly to detect them at their earliest formation would be a good tactic. But if they are more like a stream, then wading in, or possibly swimming along with them, would be a more productive approach. Much of the difference between Eastern-style mindfulness and Western-style contemplation is a consequence of choosing either the first or the second of these alternatives.

Explicit metaphors can at times introduce intellectual breakthroughs of extraordinary impact. Consider the notion that DNA "codes" for metabolic production. As we noted in an earlier discussion, this began with the realization DNA itself was made up of four nucleotides that, when strung together in triplets, generate sixty-four distinct sequences. Each of these sequences acts like a code in that it creates a specific response from the molecules that manufacture materials inside a living cell. The key point is, it would be almost impossible to understand what is going on without invoking the metaphors of *computing* and *manufacturing*. The complexity of metabolic processes is so bizarre, you need a metaphor to cut through the noise and detect the signal.

The same holds true for Darwinian evolution. Without the metaphor of farmers selectively breeding domestic crops and animals, combined with the insight that biological competitions are inherently selective in their outcomes, the idea that natural selection would both spontaneously and inevitably produce ever-increasing manifestations of complexity would never have occurred to Darwin. Conversely, once such a metaphor has been presented, its insights often seem obvious. That's what prompted Darwin's chief supporter, Thomas Huxley, to comment, "How extremely stupid not to have thought of that!"

Metaphors with such vast explanatory range and impact are rare indeed. But metaphor per se is so commonplace as to go largely unnoticed, and this is a big miss. Virtually all problem-solving statements gain power at least in part through the tacit use of metaphor. One educator wants to open her students' eyes, another wants to fill their minds with knowledge, while a third wants to challenge their capabilities. Eye-opening, mind-filling, capabilities-challenging: Can you see how different these three classrooms will be?

The single most important role of natural language analytics, once an argument has been tested for procedural validity via methods like the Toulmin model, is to surface the implicit metaphors that provide clarity, coherence, credibility, or comprehensiveness. The more alert and sensitive we become to this level of discourse, the greater our appreciation for why products of natural language find it almost impossible to build atop one another. Even works from the same author, even chapters from the same book, even paragraphs from the same essay can be grounded in different metaphors. Sometimes these metaphors cross-connect, but such junctions are never perfect, not unless they share a common "root metaphor."

Metaphors, in short, do not intermarry well. They can never participate in the secure interlocks of mathematical connections. The best we can do is

1. appreciate the creative capability of metaphor to bring to mind insights not previously considered, and

2. analyze and critique the application of any such metaphor to the situation at hand.

Each work of natural language analytics, in other words, must stand alone. It can be part of a community, a school of thought, but, unlike mathematics, it cannot reliably extend another's work unless it is scrupulously attentive to maintaining consistency of metaphor.

To conclude, within analytics we have two powerful traditions, one grounded in mathematics, the other in natural language. Both can help us critique the value of narratives as strategies for living. In addition, both types of analytics can be applied to themselves. That is, one can have analytics of analytics, but here once again the twin paths diverge. Mathematics naturally incorporates analytics of analytics as a method for systematically extending its reach. Algebra extends geometry, just as calculus extends algebra. The edifice continues to rise.

Not so with natural language. Here, when one applies analytics to someone else's analysis, one must cope with a myriad of second-order effects that inevitably arise from differing human perspectives. This is why work in the sciences, based on mathematics, appears linear, with each innovation building on the last, while work in philosophy, for example, can appear circular, with debates on fundamental issues never reaching resolution. At one remove, this can still be rewarding provided one has the patience to sort through all the nuances. At two removes, it becomes virtually impossible to cope with all the possibilities and still do justice to the topic. Sadly, an enormous amount of cultural discourse operates at two or more removes. That is, it consists of an analysis of someone else's analysis of someone else's analysis—most frequently encountered either in the aridity of hyper-specialized scholarship or the insanity of politicized debates. The best rule of thumb here is get as close to the original underlying narrative as you possibly can and keep your attention focused there.

This observation brings us to the final transition in our staircase model. For the staircase itself is a root metaphor for a kind of uber-narrative in its own right, and as such is deserving of its own uber-analytic. That is the job of *theory*.

Stair 11: Theory

One Meme to Rule Them All

By placing theory at the top of the staircase, I am not referring to the theory of just anything—I mean the theory of everything! For that, in effect, is what metaphysics claims to be—the ultimate ground for understanding everything else.

So what would a theory of everything consist of? In the terms of our staircase, it is an act of language, incorporating both narrative and analytics, that seeks to explain, well, everything. Actually, if you add no other criteria, coming up with a theory of everything is not that hard. You could say, for example:

- The universe is run by spirits who animate all things and who can be negotiated with in order to accomplish one's desired ends. Name all the spirits, figure out what each one wants, and you understand everything.
- The universe consists of four basic forces—electromagnetism, the strong nuclear force, the weak nuclear force, and gravity—and whatever is not covered by physics is an epiphenomenon of no abiding consequence.
- The universe is a dream, a projection of the imagination and has no reality beyond consciousness itself.
- The universe is lots of different things, perhaps like a supermarket, and it needs a bag full of theories to account for it properly.

Really, the possibilities are endless.

For a theory of everything to be taken seriously, however, it must measure up well against at least four standards:

1. *Precision.* It must identify and discriminate everything from everything else in ways that are stable and verifiable. (This is where the dream theory falls short.)

2. *Scope.* It must leave nothing out, nor declare any phenomenon unworthy of consideration or unreal. (This is where the four-forces theory falls short.)

3. *Structure.* It must anticipate and imply phenomena that are as yet undiscovered. (This is where the spirits theory falls short.)

4. *Coherence.* It must communicate a unity that underlies all variety. (This is where the bag of theories theory falls short.)

This set of standards is derived from the best book on metaphysics I know, Stephen Pepper's *World Hypotheses*, published back in 1942. Pepper doesn't actually present a theory of everything of his own. Rather, he provides a framework for understanding and critiquing any given candidate theory, something he calls a world hypothesis. His seminal idea is that all world hypotheses achieve coherence through the same mechanism, namely a *root metaphor,* and then compete with one another on scope, precision, and structure.

A root metaphor, as we have discussed, functions as a grand analogy, allowing one to decipher the mysteries of A by using the structure of B as a kind of decoder ring. As long as the categories used to describe A are derived directly from those that make up B, the theory will be coherent to the degree that B itself is a single, whole thing. By contrast, if the theory fills itself out with some extra categories that do not map back directly to B, then it will be incoherent and thus invalidated. Presuming it can meet the criterion of coherence, then the remaining criteria of precision, scope, and structure will evaluate how effectively the set of categories known from B are able to cope with the immensity and complexity of understanding A.

In this context Pepper identifies four root metaphors that have proved "relatively adequate" (his term) for describing how all of reality works:

1. *The universe is an intelligently created artifact.* Mind imposes form upon matter, something that lasts for a time, after which matter falls back into formless decay. Form itself, however, is eternal, and understanding it unlocks the mysteries of the universe. This root metaphor underlies Platonism, Aristotelianism, and Christianity

and is generally associated with *idealism*, or what Pepper calls *Formism*.

2. *The universe is a vast, impersonal machine.* It is governed by the laws of physics, in which all phenomena can be explained mathematically in terms of some combination of elementary material forces. This is the root metaphor popularized by Descartes, one that splits reality into two distinct spheres—mind and body—a dualism that persists to this very day. It is generally associated with *materialism*, or what Pepper calls *Mechanism*.

3. *The universe is a river flowing through time.* It is a spontaneous unfolding of intermingling currents synthesized uniquely and unrepeatably in the here and now. Patterns emerge, dissolve, and reappear in ways that can be experienced, understood, anticipated, and used to advantage, but never completely predicted. This is the root metaphor that underlies *pragmatism* as espoused by William James, John Dewey, and other, predominantly American, philosophers of the 19th and 20th centuries, or what Pepper calls *Contextualism*.

4. *The universe is a teleological journey.* It is a dialectical unfolding of *thesis, antithesis,* and *synthesis,* in which each synthesis becomes a next-generation thesis. The universe is always progressing to higher levels of integration, always headed toward a unifying, validating end. This pattern underlies both the idealism of Hegel and the materialism of Marx, both of whom propose teleological ends that magnetically attract our means. Pepper calls such philosophies *Organicism*.

By Pepper's reading, each of these root metaphors is sufficiently robust to subsume and thereby "explain away" the other three. Thus, no one of them can gain the upper hand over the others on any sustainable basis. Each has its own strong suit, aspects of experience and reality it handles particularly well, but each is also burdened with a correspondingly weak suit elsewhere. One would like to combine the best features of each, but that would create an incoherent, eclectic mess—or would it?

As you may have already surmised, we've been building an answer to this question throughout this book. The root metaphor of the infinite staircase, I believe, can pull this off. Formally stated, it says:

The universe is a systems hierarchy. It is a nested series of locally ordered systems, each with its own differentiating set of rules, each subsuming lower-order systems of less complexity and contributing to higher-order systems of greater complexity. It is not clear whether there is an ultimate top or bottom to this hierarchy, but to date we have been unable to observe either, and the metaphor does not require us to resolve this issue.

Systems hierarchy is a formal-sounding phrase, but in fact we are very familiar with this structure. Language, after all, is a systems hierarchy. The words on this page are made up of syllables, which are made up of letters, and they come together in phrases, which make up sentences, which make up paragraphs. All these levels have their local rules, yet they all come together to function in a systems hierarchy.

In such a hierarchy, each level can both support its own local root metaphor and at the same time participate in systems hierarchy as its global root metaphor. This is how we get to have our cake and eat it too. It's like one of those mixed-use-development complexes where you can have retail outlets on the first floor, office space in the floors above, and apartment living above that.

In our case, at the bottom of the staircase, we have been able to use the materialist root metaphors of physics, centered on entropy, to explain the mechanisms that drive the entire universe. But because emergence allows for the independence of levels, we are not trapped in that metaphor forever. Thus, when we reached the middle of the staircase, we were able to migrate to the pragmatist metaphors of evolution without losing connection to our materialist roots and without committing ourselves to either metaphor for the top of the staircase. Indeed, when I first envisioned the staircase, I assumed that either idealist or organicist metaphors would hold sway at the top. As it turned out, however, by substituting memes for genes, we ended up staying with the pragmatist root metaphor of Darwinism. The root metaphor of the systems hierarchy, in other words, gave us considerable freedom to pick and choose.

A systems hierarchy is characterized by a number of key elements, each of which plays a crucial role in helping us articulate our theory of everything:

- *Levels.* This is perhaps the most distinctive feature of systems hierarchies. Gradations in complexity are not continuous. Rather they appear in discrete separate states, like the quantum levels of the electron shells inside an atom, which are indeed a case example. We call these separations *levels*, implying the root metaphor of terraces or steps in a staircase, as distinct from slopes, slides, or ramps, the idea being that complexity sorts itself out into distinct layers rather than evolving along a gradual continuum.

- *Systems.* Here we mean the set of local operations that characterize any particular level. In human biology, for example, we discriminate between the immune system, the nervous system, the digestive system, and so on. Each has its own distinct set of local rules and is, in turn, made up of its own local subsystems, each with its own local rules, all integrating into a higher-level organism with its own rules as well.

- *Hierarchy.* This is simply the consequence of having multiple nested levels. Although the number of levels is potentially unbounded at both the top and bottom, nonetheless, they are unmistakably organized in a "vertical" orientation of subset, set, and superset. This stands in contrast to the "horizontal" orientation of a local system interoperating with its peers. Vertical hierarchy implies that any given system can always be both decomposed into constituent subsystems and also integrated with other systems to create a higher-level supra-system, thereby giving the root metaphor considerable predictive power.

- *Emergence.* This is the most magical property of any systems hierarchy. It refers to the spontaneous appearance at each level of a new form of complexity with a unique set of local rules. These rules clearly have dependencies on the rules and operations at lower levels, but, just as clearly, they can neither be predicted from, nor reduced to, those rule sets. Instead, at each new level of a systems hierarchy, genuinely novel phenomena appear for the first time.

- *Context.* Higher-level systems create the context for valuing lower-level systems. This applies specifically to those lower-level operations that impinge upon the higher ones. Thus, a mutation in a single nucleotide of DNA becomes significant if it alters the production of a protein necessary for the health of its host organism. Likewise, a comma misplaced in a sentence can alter the meaning and value of a legal contract. Value is assigned by higher-level systems identifying a dependency on lower-level operations.

Taken together, these five concepts—levels, systems, hierarchy, emergence, and context—represent the decoding capabilities of our root metaphor. By claiming that the universe is a systems hierarchy, we are committing ourselves to explaining every known phenomenon in relation to these core concepts.

Specifically, looking back over the three chapters that make up our efforts to date, here are the metaphysical claims we have made:

- The universe first and foremost is a mechanistic construct. Physics, chemistry, and biology would explain everything were it not for the phenomenon of consciousness.
- Consciousness emerges under the influence of desire. Its subsequent evolution is shaped by natural and sexual selection, where success is determined on purely pragmatic grounds, first at the individual level, then, with the emergence of mammals, at the social level as well.
- Values and culture are extensions of consciousness that grow out of socialization. They have their roots in our mammalian heritage, emerging prior to language, specifically via the nurture of the young and the governance of the tribe. These practices also evolved via Darwinian selection.
- Language fabricates, manipulates, and communicates memes developed via culture. Memes play the same role in the evolution of ideas as genes do in the evolution of organisms. Both encode strategies for living that are shaped by natural and sexual selection.
- Narrative and analytics are the twin engines of strategic thinking, with the humanities and social sciences foregrounding the former, the STEM disciplines the latter. All human strategies for living are anchored on these two stairs.
- Theory is a fusion of narrative and analytics that seeks to synthesize as much explanatory and predictive power as possible. The value of

the systems hierarchy model is that it both predicts and accommo-
dates emergent additions while maintaining continuity with existing working subsystems.

In developing these points, our argument divided itself into three
chapters. The first was completely materialist. It claimed that entropy,
in the sense of the universe's unceasing tendency to disperse energy as
heat, is the driving force behind the emergence of complexity. Other
forces may accelerate, brake, or steer the course of physical events, but
entropy itself is the fuel and motor driving them. Everything that exists
is an artifact of the Big Bang seeking to cool down.

Next we took on the emergence of dynamic living organisms. Here
we were pragmatist in our orientation, explaining complexity as the
consequence of competition selecting for whatever works best. Chapter
three traced an evolutionary arc from desire through consciousness to
values and culture. This is the domain of Darwinism. Natural and sexual selection represent the ultimate in pragmatism since both are driven
by the "whatever works" principle. Entropy-driven emergence creates a
cornucopia of variations from which Darwinian selection extracts the
fittest. That's what connects chapter two to chapter three.

Chapter four was a bit of a surprise. The domains of language, narrative, analytics, and theory would seem custom-made for an idealist approach, one where order is imposed from above. Each, after all,
is characterized by using form to organize material. Instead, however,
struck by the fact that both *genes* and *memes* transmit strategies for living, we extended the pragmatist phenomenon of natural and sexual
selection from biology into the realm of ideas. That took us all the way
to the top of the staircase.

The essence of our metaphysics, the claim that connects all three
chapters, boils down to this:

With respect to life on Earth, the mechanism of entropy, initiated by the Big Bang and augmented by the oscillating impact
of the sun's energy, drives the emergence of living systems of
increasing complexity. As these systems expand their reach,
they eventually encounter or create conditions of scarcity.

Scarcity initiates a competition for resources that triggers the processes of natural and sexual selection. Competitive selection favors complex systems, leading to the emergence of higher and higher levels of complexity. The result at any given point in time is a world organized around a systems hierarchy.

Systems hierarchies have no discernible top or bottom. Thus, no matter how deeply we penetrate into the elements of matter, we can never with confidence say we have reached the bottom of things. And no matter how far we extend our view out into the reaches of space, we can never with confidence say we have reached its outmost limits. Interestingly, however, we can say that the universe as we know it does have a discernible beginning (the Big Bang) and a discernible end (heat death) although both are so distant in time as to be effectively irrelevant.

So what are we to make of this world hypothesis? If we assess it in terms of Pepper's criteria, we can say:

- It has *precision*, for at each level we can develop frameworks specifically adapted to its local dynamics.
- It has *scope*, for we can stack up as many levels as necessary to accommodate the variety of phenomena we encounter.
- It has *structure*, specifically a hierarchy of nested levels, which is both descriptive and predictive.
- It has *coherence*, enabled by the root metaphor of a staircase, so that it all fits together.

By all rights, we should be able to declare victory and be done. I have to say, however, that my own reaction is more muted. I see nothing wrong with this world hypothesis—indeed, it is as "right" as I know how to make it—but it is lacking an essential dimension. It does not, as currently framed, deliver what we most want from metaphysics: a foundation upon which to build an ethical life.

There are two things metaphysics must do to meet this need. The first is to identify a personal source of moral energy to sustain and refresh us in times of trial. We call such energy *spiritual,* and metaphysics is responsible for locating it and pointing the way to accessing it. Our staircase, as currently framed, does neither. Where are we supposed to go for spiritual experience and moral support?

Second, whatever ethics we choose to commit to, our metaphysics must give them sufficient authority to withstand challenges to their validity. Religion has traditionally provided such authority through the indisputability of its sacred texts. A secular worldview, however, does not offer that option. We do have secular narratives that can explain and validate our ethics, but they do not confer authority upon them. We can also have a social consensus that authorizes our ethics, but only up to the boundaries of our social group. Another social group with an opposing consensus can deny that our ethics have any authority. From whence, then, can secular ethics derive any sustainable authority? Again, our staircase does not provide an answer.

Where can we go from here?

As a point of departure, let me note that, as an American, I am heir to two indigenous philosophical traditions: pragmatism and transcendentalism. The systems hierarchy world hypothesis has leaned heavily on the first of these two, but it has completely ignored the second. If we turn our attention to transcendentalism, we see it does locate an accessible source of moral energy in what it calls Being. It claims that Being is not only universal and omnipresent, it is also inherently life-supporting. As such, it is well positioned to authorize ethical values. If we could find a way to add Being to our staircase, we would have a bridge to connect our ethics to our metaphysics.

However, in order to keep faith with our secular approach, whatever claims we make for Being would have to be objectively verifiable. This represents a serious challenge. How close we can come to meeting it is the business of the next chapter.

CHAPTER 5

Being

A Bridge to Ethics

Historically, the relationship between metaphysics and ethics has been anchored in religion. God is presented as supreme and beneficent, and humankind is instructed to act in accordance with God's guidance. That guidance is provided through sacred texts, supplemented by expert advice, and reinforced through social norms. Success in following this path leads to happiness, joy, and eternal life. Failure results in isolation and punishment.

Of course, any actual religion is considerably more subtle and nuanced than that, but the point is, in all religions, the bridge between metaphysics and ethics is established through a uniquely privileged narrative and secured by a wealth of commentary. Such a path, however, is not available to anyone adopting a secular worldview. In that context, there are no sacred texts, and all commentary is relative. If we are going to connect metaphysics with ethics, we need to take a different path. Our path will be anchored in personal spiritual experience.

Most of us have had such experiences at some point in our lives. They may have come in a moment of reflection, or from the sight of an awe-inspiring landscape, or from a child's innocent expression of love. They may be due to something we just read, or to prayer, or meditation, or yoga, or just taking a walk on a beautiful day. Such moments touch the core of our being, what we are likely to call our soul, regardless of our religious beliefs. They center us, they refresh our spirit, they renew in us the moral energy we need to cope with life's challenges. Unfortunately, they are not always repeatable, and we need them to be so.

There are secular traditions that do seek to anchor themselves in repeatable, personal spiritual experience. One is American transcendentalism as developed initially by Ralph Waldo Emerson, Henry David Thoreau, and Walt Whitman. This tradition connects metaphysics to ethics via a radical interpretation of the idea of *being*.

Normally, we think of being as simple existence, without any active properties, as in: *John is a man. A cheetah is swift. That is a tree.* The verb *to be* is acting like a bit of sticky tape attaching an attribute to a subject. It translates roughly into "participates in the set of." Transcendentalism proposes that there is another dimension to *being* that we are missing here, what it sometimes refers to as *Being* with a capital B. Being in this context is a field that underlies all of nature, including our own nature, suffusing it with life-supporting energy. If this kind of Being could be incorporated into our infinite staircase, and if it could in fact deliver the kind of refreshing spiritual experiences we need, then we would indeed have a bridge between metaphysics and ethics.

To test such a claim, therefore, we need to address those two issues:

1. Can a field or a force like Being be accommodated within our systems hierarchy world hypothesis?
2. If so, then can its spiritual support be validated empirically without recourse to religious belief?

With respect to the first issue, the staircase framework offers at least two possible entry points for incorporating Being into our model of reality. One is the experience of *joy*, which we located on the stairstep of values. Joy produces the spiritual refreshment necessary to energize and sustain ethical behavior. For most of us, however, joy occurs spontaneously and is normally evanescent, making it unreliable as a source of spiritual support. If, however, we could somehow systematically tap into joy on a recurrent basis, that would be a different matter. Transcendentalism claims that joy is an inherent property of Being and that, through meditation or a comparable practice, we can indeed tap in to it at will. In the terms of our staircase, the claim is that consciousness, motivated by desire, can experience the nurturing value of joy anytime, anywhere. This is an empirical claim we can test.

The second potential opening for introducing Being into the staircase framework is that, by our own admission, neither the bottom nor the top of the staircase is directly observable. Transcendentalism asserts that Being is located at the very bottom of our staircase. In that position, it operates like an active field, comparable to electromagnetism or gravity, from which it can support all living things. This is not a claim we can validate. What we can say, however, is that the metaphor of the infinite staircase does not invalidate it. That means it is possible to subscribe to both the systems hierarchy and transcendentalism without internal contradiction.

Of course, just because we can subscribe to transcendentalism's understanding of Being does not mean we should. If we are going to incorporate it into our secular world hypothesis, here are the claims I believe we must validate:

1. Spiritual support for living ethically, the kind of support traditionally delivered by religion to the faithful, exists independent of our beliefs, whatever those may be.
2. This support is integrated into the structure of life itself—it is omnipresent and always accessible.
3. We can consciously experience a direct connection to this field of support at will.
4. This connection is available to everyone, again regardless of beliefs.

We will take on these claims one at time, in reverse order, beginning with the last one:

> The connection to Being that provides spiritual support on a
> repeatable basis is accessible to everyone, regardless of our
> beliefs.

In full disclosure, I am personally convinced that this claim is true based on my having practiced Transcendental Meditation for over fifty years. But that is not data so much as testimony. How, then, might we go about validating this claim?

First of all, many people practice TM, and many more practice mindfulness or a similar discipline. Both populations report they gain access to the kind of spiritual support we are talking about. That is, they experience a kind of inner peace that centers the personality in a calm state of security, thereby providing a stable foundation for ethical living. Moreover, there is a large body of research that documents physiological, cognitive, emotional, and therapeutic effects consistent with these claims of spiritual support, and it is all accessible over the Internet.

Second, despite the traditional view of spiritual enlightenment as something reserved for the exceptional few, there is nothing to indicate that these practices are elitist in any sense. The kind of meditation we are referencing here is effortless in the most radical sense, along the lines of this passage from Andrew Marvell's poem "The Garden":

> Meanwhile the mind from pleasure less
> Withdraws into its happiness . . .
> Annihilating all that's made
> To a green thought in a green shade.

The only action involved here is a redirection of awareness. It involves neither intellect nor skill. It is not like Zen meditation, aikido, or yoga. You cannot become more adept at it. All you can do is become more habitual in performing it.

That possibility is enabled by our next claim:

> We can consciously experience a direct connection to this
> field of support at will.

Again, the kind of conscious experience we are talking about has long been a part of our cultural heritage, expressed, for example, in the traditional Christian benediction:

> May the peace which passes all understanding
> Fill your hearts and minds
> And be with you now and forevermore.

The ability to *directly connect* with this experience, however, has not been widely shared. In the transcendentalist tradition, such a connection is not delivered via the intellect. In the context of the infinite staircase, it does not draw upon the higher faculties of language, narrative, analytics, or theory. Instead, it represents our animal consciousness, motivated by desire, accessing and experiencing value in a direct, personal, and unmediated way. In other words, we are smack-dab in the middle of the staircase, solidly below its upper tiers. This is a seminal insight: spiritual experience comes from the middle of the staircase, not from the top.

Nothing transcendental can be experienced once language has been introduced. Language, and everything that is built upon it, is inescapably dualistic. Robert Frost captures this idea beautifully in a little poem called "The Rose Family":

> The rose is a rose,
> And was always a rose.
> But the theory now goes
> That the apple's a rose,
> And the pear is, and so's
> The plum, I suppose.
> The dear only knows
> What will next prove a rose.
> You, of course, are a rose—
> But were always a rose.

As Frost teasingly shows, language and analytics chop up the world into categories and classes, yet underneath it all, an abiding substrate persists. When we seek Being through the medium of analytics, we get caught up in its categories and never get to the substrate. Frustrated by these efforts, we conclude that spiritual experience is not easily accessible, that achieving it is something only the elite can accomplish.

The experience of mindfulness, however, is different. It is based on desire coupled to awareness, not to intellect nor to will. We are fully conscious but neither analyzing nor concentrating, just experiencing something as simple as our breathing or a mantra. The object of experience is intended to be so ordinary, so commonplace, as to fade away, leaving us aware of just being—or, as the transcendentalist would say, of just Being.

In this context, the next claim to validate is virtually self-evident:

> Spiritual support is integrated into the structure of life itself—it is omnipresent and always available.

To be perfectly clear, lowercase *being* is integrated into the structure of matter and energy, whereas uppercase *Being* is integrated into awareness. Awareness is an artifact of consciousness. For any conscious organism, therefore, Being is integrated into the structure of life itself and is omnipresent.

To say that it is always available, however, does not mean that everyone succeeds in gaining access to it. Being can be experienced only in the present moment. If we fail to bring our attention into the present, we cannot become aware of Being. But if we do bring our attention into the present, and if we allow ourselves to experience awareness per se, then Being is indeed unavoidable. It is always there. The salient question, then, is, *Can we be there with it?*

Sometimes the answer to that question is no. In dire circumstances, any number of traumas could be the cause. Recovery from trauma can be deeply challenging because the very receptors of awareness have been damaged. It is still possible to experience spiritual support, but here conventional religion has a much stronger offer to make. People recovering from trauma need not only spiritual but social support, and church communities of all faiths excel at this sort of thing. Moreover, traumatized individuals need to bring the power of narrative and analytics to repair their psyches, and religion excels at bringing these forces to bear as well. A secular worldview need not embrace this path, but it would be foolish to repudiate it.

Even in the context of leading a normal and happy life, one can still lose one's way. The transition from childhood to adolescence, for example, can bring changes to consciousness that isolate us from experiencing spiritual support. The personality becomes deeply immersed in the upper levels of the staircase. We seek out new language, experiment with new narratives, marshal new analytics, and explore new theories. In this welter of conflicting ideas and emotions, the ability, even the desire, to connect to Being is easily lost.

Moreover, it can stay lost when we transition from adolescence to adulthood. As we take on a full complement of adult commitments, we may become so immersed in them that we do not find the time for spiritual refreshment. We may decide instead that spiritual life is not really a part of adult affairs. Without rejecting any of these things explicitly, we can neglect them by default, especially in an era of digital distractions and 24-7 work habits. Spiritual support may be omnipresent, but experience shows it is not unavoidable.

In this context, as long as life is proceeding within expected norms, we appear to get along quite manageably. Sooner or later, however, life challenges us more deeply. Then, just when we most need to experience spiritual support, we find ourselves out in the cold.

It is at this point that the very first of the four claims, the one we have left to last to address, becomes deeply important. Let's pose that claim as a question, as that is how it is likely to be experienced in this situation:

> Does spiritual support for life on Earth, the kind of support traditionally delivered by religion to the faithful, really exist independent of our beliefs, whatever those beliefs may be?

When we have run out of love, is there some external source that can replenish us? Where would such love come from? Where does any love come from? Is it somehow inherent in Being itself, or is it simply an epiphenomenon, a hormonal stimulus selected by evolution to support increases in complexity, the inexorable working out of an entropy-driven universe seeking to return to its lowest available energy state?

Actually, there is room to believe it is both. In discussing the metaphysics of entropy, we hypothesized that Earth has an interest in maximizing complexity as a means of shedding the sun's energy. Such an imperative does account for a staircase in which higher and higher orders of complexity emerge, each characterized by greater entropy-generating capabilities. Biological life represents a major step up in this regard, with humanity achieving the highest step to date. In this context, all that we value about ourselves, including our most profound emotions and dearest accomplishments, are indeed just epiphenomena. This is the materialist point of view, and it is perfectly sound as far as it goes.

The materialist account of reality falters with the emergence of consciousness. Yes, consciousness is a tool for creating more entropy, but that is not its only function. It is also a mechanism for sensing and managing homeostasis, the tendency of all living things to seek a life-sustaining equilibrium. In the realm of animal behavior, homeostasis results in what we called the Darwinian mean, a pragmatically determined equilibrium between risk and reward based on balancing desire with fear, one that maximizes an organism's chances of survival and reproduction. In the realm of human behavior, homeostasis is more closely aligned with the Aristotelian mean, a psychologically determined equilibrium among a variety of conflicting feelings that maximizes our organism's sense of well-being.

Reliable access to well-being is the kind of spiritual support we are looking for. Aristotle's strategy for securing it was through a conscious commitment to the virtue of temperance. That can work for personalities that are already in balance and have spiritual energy to expend, but it does not work for those that are disaffected, wounded, or in need. The latter need an influx of energy from outside themselves, something the Aristotelian mean does not provide for. Religion, by contrast, does so handsomely, which helps account for why it displaced philosophy at the core of the Western cultural tradition. But can a secular tradition supply a comparable spiritual support from Being?

We have reliable evidence that mindfulness and meditation, as well as other related practices, do confer a state of well-being. Earlier, we associated this state with joyfulness, but now we need to be more precise. We experience joy in two modes. One is as a sharp pang that can completely overwhelm us, the other a serene centeredness that suffuses quietly into our consciousness. The former is not a repeatable experience. Its epiphanies are good for shaking us up, waking us up, and taking us up to a higher plane. They make us aware of feelings we did not know we could have. But they pass, often quite quickly, and they leave us not only breathless but also at a bit of a loss. By contrast, the quieter mode of joy that suffuses into our consciousness is indeed a repeatable experience, one that becomes increasingly accessible through the practices we have been referencing. This is the joy that provides spiritual support, the one that underlies spiritual homeostasis.

Where, then, does such joy come from? It is important to note that it does not come from the practices themselves. What they deliver instead is an awareness that this quieter mode of joy was there all along. Once we are sufficiently aware of it, we can access the experience between practices. It can become abiding. This is the point at which *being* becomes *Being*. Note that we are not at the bottom of our staircase. There, being truly is just being. It is with the emergence of consciousness that being becomes Being. In other words, in the systems hierarchy worldview, *Being* emerges out of *being* halfway up the staircase.

This is a departure from the traditional understanding of transcendentalism, which aligns it with idealism. In that context, Being is positioned as a divine field from which reality emanates. Such a narrative may be compelling, but there is no way to verify it. By contrast, working within the systems hierarchy model, we are aligning transcendentalism with pragmatism. Our focus, therefore, must be on verifiable outcomes in the here and now. In that context, Being is verifiable only with the advent of consciousness. It could exist prior to consciousness, it could even be the foundation out of which the Big Bang emerged, but we have no way of verifying that. What we can verify is that Being provides spiritual support for consciousness and thus for everything above it—values, culture, language, narrative, analytics, theory. This means, among other things, that it is available to support and authorize ethics. That is what makes Being a bridge between metaphysics and ethics.

With that thought in mind, let us review the four standards for spiritual support we set for Being to meet:

1. *Spiritual support for human life on Earth, the kind of support traditionally delivered by religion to the faithful, exists independent of our beliefs, whatever they may be.* We have argued that conscious awareness of Being provides the kind of spiritual support traditionally delivered by religion. It does not require belief because it does not depend upon any acts of language, narrative, analytics, or theory.

2. *This support is naturally integrated into the structure of life itself—it is omnipresent and in effect unavoidable.* We have argued that access to Being through awareness is an inherent capacity of consciousness that requires no skill to exercise. Thus, for conscious beings, spiritual support is omnipresent. We drew the line at unavoidable, however, since, despite being available to all, many people still fail to secure this support for themselves.

3. *We can consciously experience a direct connection to this field of support.* We have argued that there is ample testimony, as well as scientific evidence, to warrant the claim that meditation, mindfulness, and related practices all provide conscious experience of Being as a source of spiritual support.

4. *This connection is accessible and available to everyone, again regardless of our beliefs.* The centering disciplines mentioned above do not involve committing to a religious narrative. They do not preclude narrative or religion; they simply do not mandate them.

As we turn our attention to ethics, we should keep in mind that doing good is not for the faint of heart, nor is it for the weak or tired. We need spiritual strength to do good on any consistent basis, particularly in light of the headwinds life often sends our way. We cannot impact fate or chance, we cannot change the hand we have been dealt, but we can change the way we play that hand. We can tap in to spiritual support to become more ethically engaged and morally responsible. With that thought in mind, let us now take a closer look at what it might entail.

PART TWO

Ethics

CHAPTER 6

Making the Turn

We have spent the first two-thirds of this book on metaphysics, seeking as complete an answer as we could muster to the question *What is going on?* Let us call that effort "setting the scene." This last third contains all the action. That is, given the universe as we think we understand it, and given that we are alive at this time but not forever, what are we supposed to do, and how should we go about doing it?

Ethics proposes that we should "do good." Bringing that into sharper focus is the job of the next three chapters, as follows:

- "Understanding Goodness" tackles the challenge of what *is* good, what parts of goodness belong to ethics, how and where ethics emerge in the staircase, and how narrative and analytics are used to authorize them.
- "Honoring the Ego" examines the *psychology* of ethics, focusing on the role of the ego as our moral agent and how it engages with the rest of our personality in making ethical decisions and taking action.
- With these two chapters as context, "Doing Good" focuses directly on what acting ethically actually *entails*. Here we explore the different norms and standards that apply depending on whether the

domain of action is personal or societal and whether it involves members of our community or others outside it.

Finally, we are going to close with a chapter titled "Being Mortal." Mortality, like immortality, sets the ultimate context for ethics. Whereas belief in immortality typically implies an ethic of ultimate obedience, belief in mortality typically aligns with a journey of self-realization. Ethics, in other words, end up defining who we are, not just to ourselves but to those around us. How much importance we attach to self-realization changes as we age, often intensifying as we get closer to our own end. In that context it is fitting we give mortality the last word.

CHAPTER 7

Understanding Goodness

If we are going to do good, as ethics asks us to do, we should get clear about what goodness is. One might ask, nonetheless, do we really have to spend a lot of cycles on this topic? Don't we all pretty much know the difference between good and bad? Helping somebody get back on their feet is good. Pushing them down in the first place is bad. Such intuitions ground our moral sense. Aren't we at risk of overthinking things here?

It depends upon your point of reference. Burning people at the stake because they hold heretical beliefs, or blowing yourself up with a bomb in order to destroy perceived enemies of your faith, do not seem like examples of doing good—yet both have been endorsed by the highest of authorities. Elizabeth I's Ireland, Ferdinand II's Inquisition, Robespierre's France, Mao's China, Stalin's Russia, Hitler's Germany, America's Manifest Destiny—each of these regimes endorsed sustained acts of genocide in the name of a greater good. Of course, one can see corrupt motives lurking beneath these policies, but set those aside for the moment. Instead, register the fact that some of the people doing these things actually believed they were doing good. How is that even possible?

Sadly, it is not only possible, it is virtually inevitable. It comes to pass whenever you construct a staircase from the top down instead of from the bottom up. This is what the two great traditions of Western culture—Judeo-Christian theology and Platonic idealism—both do. The problem is, if you begin with God or any other form of absolute perfection, there is nowhere to go but down. All values exist in their perfect state at the top of the staircase and become increasingly diluted or compromised as one steps down through the various levels. By the time you get to our mammalian instincts, far from them being perceived as sources of virtue, they are positioned as moral liabilities, the focus being on how best to control, curtail, suppress, or even repudiate them.

But it is these very same mammalian attributes that represent our common bonds with other human beings. The further up the staircase we go from them, the greater the divisiveness. Differing cultures generate separate languages leading to unique narratives, each explicated by its own specialized analytics, all leading to competing theories. If goodness is to be sought in theory, it is perched on the most precarious and least ecumenical spot in the whole hierarchy. Even the slightest distortion in interpretation will send shudders all the way down.

Now, theoretically these distortions should be detected by experiencing bad outcomes at lower steps in the hierarchy and correcting for them at that point. In top-down hierarchical systems, however, any feedback from such experiences is more likely to be invalidated. It will be attributed to imperfectly realizing the truth and falling prey to corrupting influences introduced lower down on the staircase. Moreover, there will always be people and institutions eager to leverage this kind of thinking in order to secure their own power. Thus, the very top of the staircase, what should be the place of greatest peace and goodness, is transformed into a source of monstrous behavior. Such is the origin of genocide in the name of goodness.

In this chapter we are going to take a different path. We are going to build our model of goodness from the bottom up, following a pathway of emergence, beginning with desire, proceeding step by step up the staircase, and ending with theory. At every step, there will be a feedback loop that connects back down to the steps below such that no step can declare independence from, or invalidate the contributions of, any other step. At the same time, each step will contribute new dimensions to our theory of goodness such that it will be as comprehensive as we can make it. Goodness, in other words, evolves as we move up the staircase, but no prior formulation is left behind. Each step transcends the prior frame of reference without divorcing itself from earlier stages.

This emergence-from-below agenda represents a departure from traditional Western approaches to ethics. As we have already noted, idealism and Judeo-Christian theology build the staircase from the top down, thereby generating a Great Chain of Being, with the highest value at the very top. Meanwhile, materialism pursues a reductionist quest to find truth by getting to the very bottom of things. It uses each step down to repudiate the step above it. It denies phenomenal reality to anything but the most foundational step, labeling everything else as epiphenomenal instead. In contrast to both these approaches, we are pursuing a form of naturalism we might call emergentism. Its method is to seek stable truth by working up from the bottom to get to the top of things, step by step, looking both backward and forward at every stage.

The emergentist view of goodness begins with desire. Prior to desire there is no need for goodness—there is just existence, be it at the level of physics, chemistry, or single-cell biology. With desire, however, comes seeking behavior that is reinforced whenever it culminates in fulfilled desire. Under conditions of adversarial competition and resource scarcity, desire and seeking behavior align with competitive advantage to generate evolution via natural selection. This feedback engine has driven the emergence of all visible living things. At this level of the staircase, goodness equates to life maintaining itself any which way it can.

Interestingly, with the emergence of consciousness, this definition of goodness does not change appreciably. Animals become more effective and efficient in their seeking behaviors, but prior to the emergence of nurturing or other forms of social collaboration, goodness remains a self-centric (or gene-centric) fulfillment of individual desire. That said, another form of goodness does begin to show itself, something we might call the "good trick." Good tricks are behaviors that create competitive advantage in the contests of natural and sexual selection—hence they are selected for and conserved wherever possible. They are not morally good. They are technically good. Consciousness dramatically increases the ability for an organism to learn good tricks through observation and imitation. This is the mimesis that drives the development of culture. We will have more to say about it in a moment. The point here is that, with consciousness, we see the emergence of a new dimension of the concept of goodness, namely *fitness for purpose.*

Goodness takes on a moral dimension and becomes part of ethics—indeed, the foundation of ethics—when we transition to the stairstep of values. Values emerge, again primarily among mammals, through social interaction—initially with one's mother, subsequently with one's father, then with one's siblings, ultimately with the local community. Each of these domains encourages different dimensions of ethical goodness, in large part because each domain is anchored by a different archetype—mother, father, sibling, and friend.

To avoid any misunderstanding, note that the key concept here is *archetype,* not *stereotype.* Maternal love as an archetype, for example, can be delivered by someone of either gender, someone who may not even be part of your genetic family, and even in extreme cases, not even part of your own species. The same goes for paternal love, sibling love, and communal friendships. All represent archetypal relationships out of which values in general, and goodness in particular, are constructed. With that context in mind, we can drill down into each to see how it adds its unique dimension to ethical goodness as a whole.

Maternal love anchors goodness in unconditional acceptance, a selfless commitment to do whatever one can to help loved ones achieve their desires. It manifests itself in acts of *devotion, compassion, empathy, gratitude,* and the like. Such emotions arise spontaneously from the act of nursing and being nursed, grounding the roots of socialization in the mutual enjoyment of a received experience of generosity and gratification. This is the primal core of ethics. All subsequent discussions of goodness evolve in relation to this, even those that end up repudiating unconditional love as weak, unmanly, or sinful. There is no entry into mammalian living that does not pass through the portal of maternal love first.

This might seem obvious, but one of the oddest things about reading the canonical works of Western philosophy is that, regardless of which philosopher you read, no one appears to remember that he had a mother. Over and over these philosophers insist on *constructing* values which, in fact, they *inherited.* They seek to invent out of whole cloth that which was already gifted to them. Who taught them how to speak in the first place? Their mothers. Who listened to them patiently when they were garbling out their first attempts at ideas? Their mothers. Sadly, however, by the time they get to writing things down and getting them into print, all traces of that past have been effaced. Now it is the isolated (and virtually always male) ego self-fashioning itself out of its own intellect. The results may often be brilliant, but they are ludicrously mispositioned and, for the most part, spectacularly tedious. So, aspiring philosophers, note to self: remember your mom.

Paternal love represents a counterbalance to the unconditional acceptance of maternal love. It consists of conditional acceptance from a father figure awarded in recognition of approved behavior. In this context goodness equates to *honor, respect, virtue, discipline, duty,* and the like. Its social function is to support and secure the rights of others, especially where they conflict with one's own desires, as well as vice versa. Conditional and unconditional love each provide a check on the other, thereby creating a balanced capability of honoring others while pursuing one's own ends. This balance is foundational to natural selection at the level of the social group where innovative-collaborative behavior creates competitive advantage under conditions of adversity and scarcity.

Ethical collaboration is further developed through peer interactions with siblings and close friends. Here goodness takes on the dimensions of *fairness, reciprocity, friendship, loyalty,* and the like. These values enable networks of trust that can function independently of the hierarchy of mother and father but that nonetheless can embrace and extend their values. Such networks are key to extending the reach of collaborative competitive advantage, ultimately enabling the subsequent evolution of the economic and social systems to come. In the absence of these values, cheating and free-riding will dramatically reduce the competitive advantage of collaboration, and the social group will be selected against.

Finally, ethics gets its most extended realization in community. Here goodness is characterized by *freedom, equality, participation, accountability, integrity,* and the like. These virtues do not show up in the animal kingdom but come into being only with language, narrative, and analytics. Once present, they can enable people who do not know each other personally to interoperate peacefully and fruitfully.

At this point in the staircase, however, we can observe maternal, paternal, and family values in a prelinguistic form among higher-order animals. The presence of such observable behavior in these creatures demonstrates that our core sense of good and bad does not come from above. It is neither transcendent nor divine. Rather, it is inherent in our mammalian upbringing. That means as humans we have a shared heritage of fundamental values that is universal. To be sure, more sophisticated faculties will emerge higher in the staircase and reshape these values, and that will lead to many kinds of disagreements. But we are mistaken if we think we were the ones to invent our values in the first place. The clay's existence precedes the potter's hand.

Continuing our progress up the staircase, as we transition from values to culture, goodness expands into a second dimension, what we earlier termed *fitness for purpose*. Culture, as we have defined it, is the mechanism by which strategies for living are transmitted across generations independent of the genetic code. This is the domain of good tricks. Here something is good if it is fit for purpose. *Arete*, the Greek word for excellence, refers to this kind of goodness. It can be applied to an artifact—a good car, a good painting, a good bomb—as well as to people—a good teacher, a good doctor, even a good villain (in the literary sense). This kind of goodness, in other words, is amoral. It is a testimony to the effectiveness or efficiency of the means, independent of the value of the ends. We are in the domain of makers as opposed to doers, where the dominant concern is utility or esthetics, not ethics.

This is a critical observation for anyone looking to connect ethics with natural selection. Darwinism is first and foremost about selecting for good tricks. Culture is the medium by which we pass good tricks from generation to generation. But culture is also the medium by which we pass along moral values, which also confer competitive advantages, specifically for mammals that collaborate to survive. Somewhat confusingly, we use the word *good* to cover both domains. This can create uncertainty and ambivalence about what really constitutes *goodness*. To keep matters straight, we should acknowledge the value and legitimacy of both domains while maintaining a clear distinction between them. This becomes critical with the advent of language. Language is the ultimate good trick. We just need to make sure it doesn't trick us.

Here we can get some much-welcome help from the *Oxford English Dictionary*. It provides an excellent roadmap for tracing the evolution of our understanding of goodness as it evolved in the English language. Its first four definitions of *good* are summarily synthetic, meaning they do not discriminate among types of goodness but take it as a unified whole:

1. Of things
2. Of persons
3. Of qualities or attributes
4. Of states or purposes

But after this, the word *good* develops three separate and distinct meanings. The first of these is *ethical*, connecting back to our discussion of values:

1. Morally excellent
2. Applied to God
3. Kind, benevolent
4. Pious, devout
5. Well-behaved (of a child)

The next three are *physical* or *emotional*, connecting back to our discussion of desire:

1. Corresponding to one's desire
2. Of things that give pleasure
3. Conducive to well-being

And the last set are *utilitarian* or *esthetic*, connecting back to our discussion of culture:

1. Of an opinion or an account
2. Adapted to a proposed end
3. Fit for a given task or relationship
4. Reliable, safe
5. Adequate to the purpose

6. Of a right or a claim
7. Satisfying in quantity or degree

In sum, goodness has three dimensions that can interact with one another to create complex ethical, emotional, physical, and esthetic states, as illustrated by the following table:

	Moral Is good	Desirable Feels good	Useful Works good	Example	Category
1	Yes	Yes	Yes	I babysat my grandson while his mother slept.	good deed
2	Yes	Yes	No	He and I are still learning how to play catch.	good try
3	No	Yes	Yes	We pretended we didn't hear her calling.	good trick
4	No	Yes	No	We ate all the cookies and lied about it.	good binge
5	Yes	No	Yes	I have sworn off drinking wine for a month.	good purge
6	Yes	No	No	I am fasting to protest human rights abuses.	good sacrifice
7	No	No	Yes	I committed perjury to stay out of jail.	good move
8	No	No	No	I tried to shoplift and I got caught.	no good

With the exception of the top and bottom rows, which are unambiguously good and bad, respectively, every other row raises a question: How good is *good* in this context? That's because the three dimensions of goodness do not blend. They remain independent of one another and contend for their own point of view.

For the purposes of this chapter, however, we are primarily interested in the ethical dimension of goodness, the one most clearly isolated in row 6, the "good sacrifice." Here *good* is neither pleasurable nor useful, only moral or, more specifically, altruistic. The reason for focusing here is that the returns on goodness from the other two modes align directly with self-interest. If it feels good or it works good, we are predisposed to endorse it regardless. This is not obviously true, however, when it comes to altruism. To make a case for aligning altruism with self-interest, we need to take the next step up on our staircase.

The case for altruism is anchored in narratives, beginning with ones installed during childhood. All cultures tell moral tales to their children, independent of religion or the lack thereof. For people of my generation and demographic, these included *Cinderella, Pinocchio, Rikki-Tikki-Tavi, Babar, Tintin, Lady and the Tramp, A Christmas Carol,* and *The Wizard of Oz*, along with a host of comic books (*Superman, Batman, Mighty Mouse, Donald Duck*) with later additions including *Star Wars, The Lord of the Rings, The Hunger Games, The Avengers*, and the Harry Potter series. None of these are explicitly moral tales, but every one of them instills a sense of altruism. This is how culture passes its judgments and values along to the next generation.

What makes this medium so powerful is that we, as readers, viewers, and listeners, *identify* with the main characters. This act of imagination is involuntary, a spontaneous act of empathy that causes us to experience the events of the narrative personally. We internalize its risks, hopes, fears, and rewards, and thereby generate memories that are truly formative. This is what lies behind Sir Philip Sidney's famous defense of poetry, that it "teaches and delights and moves to virtuous action." "Teach and delight" is a phrase he inherited from Horace's *dulce et utile*, sweet and useful. What he added, and what makes narrative so central to our understanding of goodness, is "moves to virtuous action." That is, stories can not only make you *feel* good (*dulce*, delight). They can not only show you what *is* good (*utile*, teach). They can actually make you want to *do* good (move to virtuous action).

Let me give a personal example. When I was a boy growing up, TV was just coming into its own, and Westerns were all the rage—*Wyatt Earp, Maverick, Cheyenne, Sugarfoot, The Rifleman, Wanted Dead or Alive, Have Gun Will Travel*, and the like. I watched them religiously, worrying if we went out to dinner, would we be home in time to see my favorite show? I wanted to be all those heroes. In my imagination I *was* all those heroes. That's what playing cowboys and shooting bad guys was all about. All my role models were pretty much made from a common formula—set apart at the surface by some unique skill or attribute, but united in their essential goodness. They were all in service to their communities, although some operated at the edges of the law. Regardless, they always did the right thing. Countless experiences of narratives enacting this code built into me a comparable desire, one that was encouraged by my peers and parents through various kinds of signaling. The point is, this *was* my moral education. Yes, there was Sunday School, and later on, chapel and Bible readings, and always there have been great conversations about right and wrong and the like, but none had the intimacy of these connections, the direct access to my identity, the total allegiance of my mind and heart and imagination.

Narratives of goodness, in other words, set our moral compasses. Because they constitute such a powerful means to shape behavior, they inevitably get appropriated by larger master narratives to promote and sustain society's institutions of power. Over time these master narratives become corrupted as those in power use them to manipulate our values in order to achieve their desires. In Western culture, ever since the Early Modern period, writers and thinkers have sought to apply analytics to free their cultures from a legacy of corrupted narratives. During the Enlightenment, the focus was on repudiating religious dogma. This was followed by Romanticism's rebellion against social dogma, then modernism's war against political and economic dogma, culminating in postmodernism's rejection of cultural dogma. In each case, analytics were used successfully to deconstruct and expose the embedded power narratives that proliferate and sustain relationships of exploitation.

Unfortunately, in throwing out all that dirty bathwater, we threw out the baby as well. Shared master narratives connect us to our culture and values, and by so doing, authorize our ethics. Now, from a practical point of view, we in the West still default to Judeo-Christian ethics and values, but this is more a product of inertia than faith. That is, our ethics are in effect a carryover from a prior master narrative that is no longer foundational. We all believe in goodness, but no one can say for certain where it comes from. And when goodness itself becomes threatened, no one can say for certain how to protect it.

This is not a stable situation, but it is the hand we have been dealt, so we must figure out how to play it. To begin with, we should be clear that we are not questioning the existence of goodness itself, only its location. Religions all have master narratives that situate goodness at the top of the staircase. From that lofty perch they authorize good behavior because it is a mechanism for aligning with the highest power in the universe. The secular master narrative is different. The universe, driven by an unceasing increase in entropy, develops a hierarchy of complexity that on Earth includes the evolution of all living things. Within that hierarchy of living things, goodness is situated in the middle of the staircase, coming into play just after the emergence of consciousness, well before the emergence of narrative. Ethical goodness comes into existence with parental nurture and discipline. This goodness is not an artifact of narrative. It is not even an artifact of language. Rather, it is an inherent property of the mammalian life cycle, installed prior to language, and prior to narrative.

This is a critically important point because once language and narrative take the stage, we lose our universal connection to the rest of humanity. That is, because all humans are mammals, anything that is inherently mammalian is inherently universal for humanity. We are all made from the same cloth. As soon as we engage with language, however, we start to become separated from one another. There are six thousand languages extant today, so no one can communicate with everyone. Add cultural narratives, each of which divides the human community into an us and a them, and we become even more separated. We lose our universal connection to each other.

Of course, we cannot do without language or without narrative. They are fundamental to our strategies for living. Together, they account for the good, the bad, and the ugly—not to mention the beautiful, the funny, and the useful. Language-enabled narratives are what give our lives meaning. But every narrative is made up of protagonists and antagonists, and once we identify with the protagonists, we are set against the antagonists. In human affairs, this inevitably ends up pitting one community against another, a we set in opposition to a they. We can choose sides, and we can choose how we behave with respect to our antagonists, but we cannot do without them.

This takes us to the heart of the problem of authorizing ethics. Ethics are authorized by narratives—specifically, metaphysical narratives. Those who buy into the narrative buy into the ethics. But there are no universal narratives, and thus there is no universal buy-in. Ethics can only be authorized locally. Goodness, because it emerges prior to language and narrative, can be universal, but ethics cannot. There will always be ethical conflicts. They are inescapable.

Nonetheless, goodness and ethics are inseparably linked. Ethics are the mechanism by which goodness acts in the world. Without ethics, goodness is inoperable. That said, the relationship is not purely linear. Just as an ecosystem can emerge out of diverse species adapting to one another, so goodness can emerge from diverse ethical systems adapting to one another. There is no guarantee that it will, however. Indeed, this same force of emergence can also lead to evil.

Evil, to draw once again upon the *Oxford English Dictionary*, is defined first and foremost as "the antithesis of good in all its principal senses." In our secular metaphysics, goodness is not transcendental, and neither is evil. That is, because evil enters our staircase at the level of values, it does not descend from above but rather emerges from below. Specifically, it emerges from consciousness coping with thwarted desire. Desire, seeking goodness, encounters its antithesis instead. Instead of nurturing, we experience neglect and abuse; instead of discipline, cruelty and exploitation; instead of fellowship, ridicule and humiliation; instead of goodwill, suspicion and rejection.

It doesn't matter what circumstances cause these breakdowns—abuse, drugs, bullying, racism, crime, terrorism—if we survive them, we adapt. We develop our own narratives, ones that are, by necessity, dark, anchored as they are in pain, fear, and anger. We enlist others in these narratives to build our own protective community, because at the end of the day, we are all still mammals, and we still need social support. At the core of this community is a fight-or-flight, kill-or-be-killed philosophy, where learning "good tricks" is the key to the game, and altruistic goodness is a fool's move, something to exploit, never to trust.

In short, evil is a consequence of the mammalian social contract gone bad. In the animal kingdom, this outcome turns out to be more the exception than the rule. Evil does not scale in nature because outcasts cannot communicate with one another and thus cannot congregate effectively. In human society, however, there is much greater scope for groups of outcasts to take root and multiply. Given the resources of language and narrative, reinforced by cynical analytics and conspiracy theories, now further amplified by digital communications and social media, human evil can scale to truly horrific dimensions. We have seen its ravages on every continent, and every society is struggling to mount an effective response.

Part of the challenge is that to confront evil directly—often the only viable response—is to adopt its fight-or-fight, kill-or-be-killed ethic. In so doing, we undermine our connection to goodness, the very thing we are seeking to protect. There is no easy way around this problem. Among other things, it calls for altruistic self-sacrifice, putting not only our lives at risk but our values as well. Such a demanding strategy would have to have strong roots in a culture for it to be viable. How could this kind of self-sacrifice evolve under conditions of natural selection?

People who reject the idea that altruism could naturally evolve do so typically from the point of view of the *selfish gene*. Genes are selected by virtue of the survival and reproduction of the organisms that host them. How could it further the evolution of any gene-based trait if the organism sacrifices itself instead of reproducing? Biologically, it cannot. But culturally it can. That is, from the point of view of the *selfish meme*, altruism does reproduce. It spreads the word by inspiring, engaging, and enlisting others in its cause. If the meme turns out to contribute to a survival mechanism, and that mechanism supports the survival and reproduction of the species as a whole, then cultural evolution will select for it even when, or perhaps especially when, the meme spreaders sacrifice themselves in the process.

With respect to mammalian strategies for living in general, altruism, because it furthers a culture of mutualism, does indeed confer competitive advantage. Mothers will risk their lives to defend their babies. Alpha leaders will risk their lives to defend the pack. Mutualism enables mammals to develop and coordinate specialized capabilities across a population, thereby giving greater variety and scope to their strategies for living. It also amortizes survival risk across a broader population, thereby making the species as a whole more adaptive to change. These are the evolutionary returns that reward investing in self-sacrifice.

Nonetheless, mutualism has numerous vulnerabilities, and evil can capitalize on them through a variety of good tricks. It can exploit trust to gain loyalty and exploit fear to retain it. It can ignore guilt and shame as constraints on behavior. It can undermine the appeal of altruism, dismissing it as loser behavior. It can lie, and cheat, and vilify at will. It can privilege power over goodness, building upon narratives of dominance and intimidation. As our century is witnessing, these are powerful tools indeed.

Goodness has a two-pronged approach for fighting evil. Its most productive returns come from preempting the hardships and isolation that nourish a culture of evil in the first place. This means staying alert to, and empathetic with, people at risk in our community and proactively seeking out and supporting creative solutions to the challenges that are damaging their lives. This is a strategy that consumes a considerable amount of spiritual energy. It also takes time.

When there is not enough time, when energy is lacking or the path forward is irretrievably blocked, then altruistic self-sacrifice must fill the void. In the short term, the outlook for success is normally not good. The goal instead is to buy time for third parties to perceive the threat to themselves and enlist in the battle to curtail evil's further expansion. Once a perimeter has been established and stabilized, then one can revert to the earlier, more productive approach.

It is important to realize that the goal is not to destroy evil. That is impossible. The conditions from which evil emerges are endemic to life itself. Seeking to destroy it only creates more evil. The goal instead is to isolate, contain, and undermine its influence. These are practical, achievable outcomes, but they do require patience, love, and understanding, for setbacks are inevitable. Recourse to spiritual support under such circumstances is critical, hence our emphasis on securing access to Being as foundational to leading an ethical life.

Finally, while it has been important to address the existence of evil, we should not overestimate its role in ethics in general. Most of life is conducted within the confines of one's own community, where relationships of trust are normal, and goodness includes acting for the benefit of others. Ethics sustain these relationships. They give credibility to our promises, substance to our actions, and meaning to our lives. They are who we are when we are being our best selves.

CHAPTER 8

Honoring the Ego

Now that we have explored the domain of *what is good*, we are poised to turn our attention to the task of *doing good*. Before we do so, however, we first need to ask, who is this *we* we are talking about? Who exactly is going to do this good, and what do we know about them? Are they trustworthy? Are they competent?

For our ethics to operate in the world, we ourselves must act as moral agents. Fortunately, there is a part of the personality that is well adapted to this challenge. In common parlance we call it the ego. We are not using the term in any specifically Freudian sense but rather with a nod toward its meaning in Latin. *Ego* in Latin means *I*. The ego is that part of the personality that refers to itself as *I* or *me*. It is the part that talks, verbally interacting with others and using speech acts to accomplish things socially. It is the part that makes choices and acts upon them, albeit often under the influence of other parts of the personality. And there's the rub.

For the ego is not the whole of the personality. We know this not just from personal experience ("I am not myself today") but from all the narratives our culture has passed down to us, and all the analytics that have been applied to those narratives. Each era has expounded its own theory of psychology. Whether it is a theory of humors, or a battle betwixt reason, will, and appetite, or an allegory of Everyman interacting with a series of personifications, or something considerably more contemporary like Pixar's *Inside Out*, psychology has always been viewed as a communal affair. Within this mental community, various parts of ourselves each have their own say, whether or not we can verbalize or even be conscious of what they are saying. Regardless, society holds us morally accountable for our actions, and that accountability falls squarely on the ego.

It doesn't take much imagination to realize that if the ego is going to have any success in life, it—and by extension *we*—had better follow the Socratic maxim: *Know thyself.* So if we think about the ego for a moment as a kind of committee chair, one with governance responsibility for the personality as a whole, who else sits on the committee, and what interests do they represent?

This takes us fully into the domain of psychology, and who better to begin with than Sigmund Freud. In Freud's view, the personality has unconscious, preconscious, and conscious domains that he later subsumed into his version of the committee model, consisting of the id, the ego, and the superego. In this framework, the stair of desire is characterized as libido, the energy that drives all our actions and represents the domain of the id. Our desiring libido brings us into conflict with our society, which in turn forces us to repress some urges, a capability that we internalize as the superego and that we located on the stair of values. The ego in this context is an agent that negotiates life within these opposing poles, aligning with the stair of consciousness. Here's the complete picture:

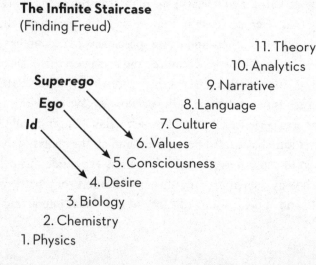

The Infinite Staircase
(Finding Freud)

```
                                            11. Theory
                                          10. Analytics
           Superego                        9. Narrative
             Ego                          8. Language
           Id                            7. Culture
                                        6. Values
                                    5. Consciousness
                               4. Desire
                           3. Biology
                        2. Chemistry
                    1. Physics
```

Freud's model got a fresh coat of paint in the 1960s when the "I'm OK, you're OK" movement reframed the core elements of id, ego, and superego as *child, adult,* and *parent.* The structural dynamics and relationships remained the same, but the whole tone of the negotiation was moved from one of warring forces to one of familial interactions, something that, if nothing else, allowed for some humor and folk wisdom to insert themselves into the situation.

One of Freud's contemporaries, first a follower and subsequently a rival, was Carl Jung. Jung broke from Freud's model to create a more dynamic taxonomy of the personality, one in which the persona plays a role comparable to the ego. *Persona* comes from the Greek word for mask, and it represents the socialized part of the self that lives up to the expectations of the social group—the ego aligned with the superego, the adult aligned with the parent. To sustain this alignment, we must repress those elements of ourselves that do not conform to society's expectations. These repressed elements coalesce in our unconscious as action figures, three of the most notable of which are:

1. The *shadow,* representing all our dark urges that are socially unacceptable (think the Incredible Hulk or Lady Macbeth),
2. The *anima* or *animus,* representing a mystical attraction to our opposite gender (think Cathy's Heathcliff or Odysseus's Circe), and
3. The *witch* or *wizard,* representing our imaginative ability to transcend time, space, and even mortality itself (think Merlin or Galadriel).

Jungian psychology portrays personal development as a series of dialectical encounters with these figures, each encounter ideally culminating in the acceptance and integration of an additional part of the total self. Thus, it is firmly aligned with the stairs of narrative and analytics. Not surprisingly, Jung's ideas had enormous influence on a generation of literary critics, most notably Northrop Frye. Following Jung's lead, Frye showed how these very same archetypes infuse literary narratives, which in turn can then be interpreted as enacting psychological journeys of discovery, encounter, and integration.

Finally, to offer a third perspective, there is a simpler rendition of the committee model that focuses simply on the relationship between the *I* and the *self*. In the Indian tradition of Vedanta, as well as the contemporary psychology of *mindfulness*, the internal dynamics of the personality are represented by the yin-yang symbol, in which a circle is divided in two by a curving line, creating two intertwined teardrops, each containing a dot of the opposite within it:

In this model, the *I* is represented by the lighter element and the *self* by the darker one. The goal of mindfulness is to expose the I to the self, to teach it self-awareness, and through that awareness to connect it with the universe as a whole. Self, in other words, ultimately merges into Being. The integrated personality engages with finite time and space via the I reaching up the staircase, and connects with the infinite via the self connecting to the very bottom of the staircase. The actual integration itself takes place on the stair of consciousness via meditation, mindfulness, or some similar practice.

In sum, we have three reasonably current psychological models (with a number of others on the bench, should we want to make a substitution) in which the personality is represented as a dynamic relationship between the ego—all models have some version of an *I*—and a company of other forces. Let us take it as a given, then, that the ego is a member of a larger set of entities that make up the personality as a whole, and move on to the question of what role it should play in the context of ethics.

With respect to ethics, the ego represents the personality as a whole in the affairs of the world and is responsible for acting out its collective values. When the ego departs from this role to pursue its parochial interests instead, bad outcomes ensue. To begin with, opposing egos get in the way. Think of Marlowe's Faustus pitted against Mephistopheles, Shakespeare's Shylock against Portia, or Star Wars' Darth Vader against his own son Luke. Even absent opposition, the things the ego pursues tend to lose their value upon being gained, as any number of post-Christmas discarded toys can bear witness. In the end, even the pursuit itself loses its luster—what one might call the "seducer's dilemma."

In such cases the ego first becomes bored, then distracted, and then disturbed. Silence becomes unnerving, evoking an absence of meaning, reminding the ego of its own vanity and, ultimately, of its own mortality. The thought of its own death is intolerable. To avoid it, self-distraction becomes compulsive. Love becomes something wanted, but not given. Life is drawn more and more into superficial engagements that fill time but do not fulfill longing. Dissatisfaction builds, and a life of quiet desperation ensues, the ego becoming outwardly passive but inwardly aggressive, always seeking chances to get back a bit of its own.

Alternatively, the ego may rebel against the ignominy of such a fate, demanding something more. Think of Tamburlaine, Dorian Gray, or the Marquis de Sade. They lash out against any person or thing in their way. They pursue dark interests. Sensation displaces emotion, which is initially a thrilling result, but eventually its allure dulls. Transgression intensifies sensation, reigniting the thrill, but only for a while. Self-loathing follows, and unlike its predecessors, it does not fade. Without redirection there is no happy outcome.

Such redirection, it turns out, is not all that complicated. The ego need only embrace one simple idea: *It's not about me.* Life is not about me. Even my own life is not about me. It includes me, I play an important role in it, but it is not *about* me. Instead, life—both all life and my life—is best understood as a narrative in which I participate.

Narrative is the medium through which the ego finds meaning and purpose. We may not have the stature of a Faustus or a Hamlet, but we too are characters in a play, one that began long before we came onstage and will go on long after we leave. Parents teach us what they know of the script, and we do the same for our children. Friends add their bits too, as do enemies. A great deal of improvisation is involved, but there are established plotlines as well, laid down by culture and reinforced by society. Our part, the ego's true role, is to play along. To do that, however, we have to learn how to get out of our own way.

A great example of an ego getting stopped in its tracks to learn just this lesson can be seen in the following (abridged) poem. It is by George Herbert, a clergyman and a contemporary of John Donne, in the generation of poets that succeeded Shakespeare. It is called "The Collar," a reference both to the clergyman's white collar and to a yoke around an oxen's neck. Herbert's ego is rebelling against the enforced humility of his Puritan faith, embracing instead the rhetoric of a privileged aristocrat. His English may seem a bit archaic, and some of his words may be unfamiliar, but listen to the tone of voice—it is as modern as could be:

THE COLLAR

I struck the board, and cry'd, No more;
   I will abroad.
 What? Shall I ever sigh and pine?
. . . .
 Is the year only lost to me?
   Have I no bays to crown it?
No flowers, no garlands gay? All blasted?
   All wasted?
 Not so, my heart: but there is fruit,
   And thou hast hands.
 Recover all thy sigh-blown age
On double pleasures . . .
. . . .
 He that forbears
  To suit and serve his need,
   Deserves his load.

But as I raved and grew more fierce and wild,
  At every word,
Methought I heard one calling, *Child*:
  And I replied, *My Lord*.

While Herbert's is clearly a Christian narrative, for our purposes it need not be. All we need to bear witness to is how an unchecked ego—frustrated, isolated, and completely self-absorbed in its own narrative—hears a voice outside itself and takes it to heart, capitulating instantly. What "The Collar" reveals so directly is that the ego, despite all its protestations to the contrary, knows it makes a poor master. It struts and frets its hour upon the stage, and then . . . it struts and frets some more. It wants the play to be about itself, but the other actors are not cooperating, and when it tries to take charge, it lacks conviction as to how to proceed, resorting to bluster instead. Needless to say, it is greatly relieved when at the end of the poem it is able to put down this burden.

On the other hand, while unable to play the master, the ego makes for a wonderful servant. Egos, at the end of the day, love to do stuff, to make an impact, to be of use. More than anything else, they want their actions to have meaning. But that meaning, that validation, must come from outside, from some externally validating narrative. It is not something the ego can create for itself. Every time the ego tries to do so, it ends up trapping itself inside a set of values and ventures that do not resonate with the outside world. Instead, for the ego to succeed, it must participate in something that confers meaning upon it. It must put itself in service to something beyond itself. What that something is will differ for each of us, but it will always manifest itself in the context of a narrative.

We cannot choose the narratives we are born into, but as adults we can choose the ones we will live by. This is the foundation of moral accountability. Ethical behavior consists in taking responsibility for the narratives we embrace and acting in accordance with the norms they imply. Right and wrong are not absolutes but are instead derived directly from the roles and narratives we have chosen.

Choosing a narrative is not optional. Without a narrative, there is no context for a strategy for living, and the ego cannot function. It simply does not know what to do. Our narratives can be organized around almost any theme: relationships—job, marriage, religion, team; achievements—graduation, buying a home, getting a promotion, raising a family; self-actualization—publishing a poem, shooting your age in golf, helping someone learn English as a second language; or any number of others. There is an enormous amount of freedom here, filtered through the values and culture of the society in which we live.

In this context, the ego chooses both the narratives to which it will commit and the kinds of roles it will try to play within them. Of course, these roles will continue to evolve, and the narratives themselves are rarely stable. Jobs can disappear; so can relationships. Life demands continual adaptation. A level of stability can come from grander narratives, ones that transcend local circumstances and give meaning and coherence to broad swaths of human endeavor. By embedding our personal narratives in a grander narrative, we can inherit some of that stability and incorporate it into our own life. That's what it means to live within a tradition, be it religious, spiritual, tribal, national, professional, or familial.

It is hard to overestimate the advantages of living within a tradition. When you are able to do so, the vast majority of ethical choices are already made for you, along lines that have already been proven reliable. The ego is free to operate without a lot of second guessing. By contrast, when individuals or societies repudiate their traditions—often with good cause, as in the case of colonial, racist, or sexist regimes, for example—the burden they take on is enormous. Now even seemingly straightforward decisions must be reexamined to determine the ethical path, something we see, for example, in contemporary debates around political correctness. Not only is the overhead burdensome, but the odds of getting things right the first time are poor. If there is no better path to take, so be it, but we must at least acknowledge it will be slow going for quite a while.

Part of what makes free ethical choices so challenging are the inevitable conflicts of interest that arise, not only with others but inside ourselves as well. Among the various internal committee members that make up our personality, whom should the ego listen to? Should we go with our gut? Our heart? Our intellect? Our conscience? We carry a myriad of voices in our heads, so it behooves us to be thoughtful as we poll them for advice.

At the end of the day, whatever choices we make, the environment will respond. In effect, ethical choices are like bets. They result in win-or-learn outcomes—either goals achieved or lessons learned (provided we survive). This is Darwinism at its core. The very force that got us to where we are is the one we must master to move forward. Ethics represent our strategy for living within a Darwinian context. Each of us has been dealt a different hand, hence the need for pluralism. We are all united nonetheless by a common goal to play our given hand as best we can.

One final thought: We should be respectful of the pressure we are putting on the ego. It is easy to criticize ourselves or others for being egocentric or egoistic or even egotistical, and there are times when such criticism is warranted. But we should always remember, the ego takes our risks for us and takes the first brunt of the consequences. We can sit back and kibitz and cluck our tongues, but life is a game that must be played in real time. There are no do-overs, and the ego must make the best of it, come what may. Things are bound to get messy sometimes, and we are going to make fools of ourselves sooner or later. That's not the ego's fault. That's the human condition.

And so, as we turn to the business of putting our ethics into action, to the realm of doing good, we should keep in mind that we are not trying to play a game of perfect. Doing good, as we have repeatedly noted, is not for the faint of heart. Often, it is not even clear what doing good means. And even when it is clear, there can be formidable obstacles in our way. We are all trying to do the best we can all the time. Our challenge is to make our best good enough.

CHAPTER 9

Doing Good

This is the culminating chapter in our effort to connect traditional Judeo-Christian ethics to contemporary secular metaphysics. It is about *doing good*. Specifically, we are concerned with the altruistic definition of goodness—*behavior that is beneficial to others*—and how we as actors, leveraging our egos, can behave ethically in the context of leading our lives. What does it take for any of us to actually *do* good?

The answer to this question, like many things in life, is contextual. Ethics is not one code of behavior but rather a set of codes from which we select depending on the situation we are in. To put ethics in perspective, then, we need to first get the lay of the land.

The Geography of Ethics

When we think about ethics, to the degree we do so at all, we tend to conjure up an abstract, uniform landscape with little or no local color, something like the blank stages popularized by the theater of the absurd. Thus, like Vladimir and Estragon in *Waiting for Godot,* we can end up feeling delocalized and disoriented, caught up in a web of abstractions, cut off from our roots. A more contoured and textured geography of ethics would be more true to life, one that can take into account the multiplicity of locales and landscapes in which we live and act.

Just as maps are organized around north and south versus east and west, so the geography of ethics I want to present is characterized by two seminal distinctions. The first is between *high-trust* and *low-trust* relationships. High trust is characterized by first- and second-person interactions, *I-you-we* relationships, that take place in a world of family and friends, colleagues and communities, teams and teammates, a context in which we align and adjust our beliefs and actions to support each other. They stand in contrast to low-trust relationships, a domain of far less intimate third-person interactions with *him*, *her*, and *them*, people we do not know and with whom therefore we cannot assume a relationship of trust. This is the world of public services, retail transactions, shared transportation, temporary colocation, or common general employment. In such venues our primary goal is to pursue our own agenda, behaving as normally as possible both to avoid drawing unwanted attention to ourselves and to allow everyone else to go about their business as freely as possible.

High-trust and low-trust relationships give rise to different personas, different ways of representing ourselves, based in large part on different levels of assumed intimacy. The world of *us* is very separate from the world of *them*. What is surprising is the degree of polarity involved here and the abruptness of the transition between the two states—there is very little middle ground in between. That is, the geography of intimacy is sharply divided between high-trust private networks and low-trust public networks such that in most real-world situations, one has to choose more or less instantaneously between two contrasting behavioral norms. You make eye contact with friends, but not with strangers, especially not in New York City. Should you give a holiday gift to the mail deliverer? Should you comment on the book title someone is reading on the bus? Often we get caught out between the two—Am I supposed to hug this person or just shake hands?—this sort of conundrum being itself a testimony to the bifurcating undercurrents in play.

Now, because we humans have been socialized as mammals—and not as reptiles—we have a marked preference for high-trust private networks, places where we can let down our guards, enjoy life, innovate, and be our best selves. As a consequence, sometimes we tend to look down on low-trust public networks as a necessary evil and something of a burden—tests to pass, forms to fill out, lines to stand in, traffic lights to tolerate. In reality, however, regulated public networks are something of a miracle, a testimony to the power of memes.

No other species has ever come close to creating such a powerful mechanism to deal with low-trust interactions. Regulation and compliance are distinctly human artifacts rooted in our unique capabilities for language, narrative, and analytics. Lacking these capabilities, other species have no means to create a buffer zone between their own high-trust in-group and the rest of the world. In nature, the lamb simply cannot lie down with the lion. In human society, there's a good chance that it can. Regulated low-trust networks allow highly distinct groups with markedly different values and cultures to inhabit a common place and interact at arm's length to pursue their own ends. Moreover, they make up a large portion of the geography of ethics—in effect, all of our public life. So, as we go forward, we should not make the mistake of discounting their value.

The second dimension that defines the geography of ethics is between *private interactions* and *public interactions*. Private interactions attach to individuals. They are one to one, or one to few, typically with people you know by name. Public interactions, by contrast, are between individuals and communities—other people again, but not necessarily ones you can name—as well as with causes, narratives, religions, nations, governments, or the like. They situate our identities in a much larger, more abstract space.

Both kinds of interaction lay claim to our ethical allegiance but in different ways. Our desire to behave in ways beneficial to others comes from experiencing the rewards of intimacy in personal relationships. When we extend that desire to a public level, we generalize our obligations across a much broader sphere of behaviors by tying the way we act to the meaning and value of our personal identity. The two spheres interoperate. That is, we end up using our societally influenced public commitments to help us navigate personal situations that have an ethical dimension, and conversely, what we learn from our personal interactions helps refine and reshape our beliefs and our faith in communal ones. Overall, our ethical identity and reputation are determined by what communal values we choose to be loyal to and how well our personal behavior reflects those commitments. To be ethical we must be accountable to both realms, each reconciled with the other.

Bringing the two pairs of dimensions together—high-trust versus low-trust social networks and personal versus public interactions—creates the geography of ethics: a total of four domains, each organized around a different core value, as illustrated by the following diagram:

The Geography of Ethics

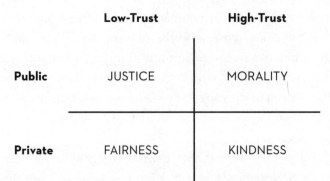

	Low-Trust	High-Trust
Public	JUSTICE	MORALITY
Private	FAIRNESS	KINDNESS

The claim of this diagram is that *doing good* is situational with respect to each of these four regions. In each case we want to see how that plays out within the specific locale in question as well as in situations that cross geographical boundaries.

In this context, our capability to do good evolves as we mature personally, mastering new skills as we ourselves move up the staircase. The journey begins with *kindness*, an intuitive, subjective impulse to act for the benefit of others, born out of an experience of empathy. This is a pre-linguistic capability that is manifested in animals as well as small children. It aligns directly with the stair we labeled *values* and represents our first foray into doing good.

The next stage in our moral evolution involves embracing the principle of *fairness*. Like kindness, fairness is intuitive, but unlike it, it is objective rather than subjective. That is, fairness requires us to see ourselves objectively as a member of a social set, each member having equal entitlements and responsibilities. Fairness asks us to support this perspective even when it is not in our immediate self-interest. To do so, we must call on the resources of *culture* and *language*, because fairness entails understanding. This puts it beyond the reach of animals but makes it accessible to children, even as early as kindergarten. Taken together, learning to be kind and to play fair make up the bulk of any child's moral development.

The next step in our evolution is a big one. We move beyond intuition and personal relationships to apply abstract reasoning to ethical choices that have communal implications. Instead of simply dealing with individual situations on a case-by-case basis, we seek to develop a set of principles and rules of conduct that can be applied on a broader basis. Here we are in the domain of *morality,* and we are relying upon the stairs of *narrative* and *analytics.* Narrative shows what good looks like in a paradigm case, from which analytics extracts ethical principles for future applications. The goal is to apply these principles across as broad a set of situations as possible as a means of maintaining integrity and accountability. We typically take up this challenge during adolescence, a time when we seek to define our moral identity relative to the social demands to which we are exposed.

Finally, the most complex domain of ethics is *justice.* Here the goal is to extend the reach of analytics beyond the subjective interpretation of communal narratives and apply it to the objective interpretation of societally determined rights and responsibilities. In modern democratic societies, this effort is founded upon *theory,* specifically a theory of the state, emerging out of an implicit social contract, translated into a political system that has executive, legislative, and judicial powers, operating under the consent of the governed. Bringing such a state into existence and maintaining it against internal and external threats is a thoroughly adult undertaking.

Altogether, then, to fully realize our ambition to do good, this is the mount we must climb:

The Infinite Staircase
(Situating Ethics)

<pre>
                                Justice ————————→ 11. Theory
                             Morality ——————→ 10. Analytics
                                          9. Narrative
                      Fairness ——————→ 8. Language
                                   7. Culture
               Kindness ——————→ 6. Values
                         5. Consciousness
                      4. Desire
                   3. Biology
                2. Chemistry
             1. Physics
</pre>

We start this journey at the base camp of kindness.

Kindness

Kindness is the quality of being friendly and considerate. It is best appreciated by its family of synonyms: *benevolence, humanity, generosity, charity, sympathy, compassion, tenderness.* What they all have in common is an emotional disposition to make the welfare of another the focus of our own behavior.

Kindness is the glue that holds the human family together. Without kindness we self-isolate, pursuing private agendas without empathy for others, without consolation for ourselves. Kindness overcomes that isolation, reconnecting us to each other, to our inner being, to our place in the world. The smallest act of kindness lights up whatever darkness surrounds it.

As an ethical principle, kindness aligns behavior with the spiritual force of love. By being kind, not only are we acting for the benefit of others, we are also letting love pass through us. That experience is cleansing, energizing, and refreshing. Whenever we are being our best selves, we are likely to be riding atop an undercurrent of love. Without that undercurrent, our acts of kindness are at risk of becoming tactical or patronizing, testimony more to our pride than any impulse to do good for others.

Kindness, for the most part, is local in time and place, the stuff of everyday living, something that wants to be "always on." This poses an existential challenge, for most of us are unable to be always on. That leads to acts of neglect or, worse, of meanness or cruelty. The darkest threat to kindness is a persistent state of anger, a suppressed rage, due perhaps to a wound born out of some prior violation of trust. Such wounds drive the personality to recurrent acts of retribution, irrespective of the target. Alcohol and drugs, sometimes taken to dampen this anger, more often have the opposite effect, and in such a case, there is no remedy that works short of a major intervention.

More commonly, however, our efforts to be kind are challenged by more mundane forces. Tiredness, be it physical, social, or spiritual, tops the list. Kindness requires an outflow of personal energy that sometimes is just not there. At such times, the best we can do is to acknowledge the obligation to be kind but beg off temporarily from the actual performance, hoping to "catch up" at some later date. In relationships of trust, we cut each other some slack, but all the while in the back of our brains there is a quid-pro-quo calculator keeping track. If the balance gets really off, we must address the issue explicitly or suffer a change in the relationship.

Kindness is the fuel that powers our most intimate relationships. Without these, life is indeed "solitary, poor, nasty, brutish, and short." Fortunately, kindness is the most accessible of virtues. It does not require access to the stairsteps of theory or analytics. It doesn't even require access to language or narrative, although both can amplify it greatly. Anchored on the stairstep of values, well before the emergence of language, kindness simply requires channeling whatever love you have received and letting the rest go. Being present for the other person, and without intending it per se, just being your best self, is a wonderful experience for both the giver and the receiver.

Given all this, one might ask, if the goal is to do good, why do we need anything more? Why not just be kind to everyone and be done with it? The challenge is that kindness implies a relationship of trust. Most of us find it hard to extend such trust outside the boundaries of our family and social group. Saints may be able to overcome this barrier, but for the rest of us, it is a high bar. Ordinary everyday kindness, in other words, does not scale. Additionally, it has a problem of scope. That is, there are moral issues that kindness simply does not address, including the ones around fairness, morality, and justice that we are about to consider. Nonetheless, all things considered, we take our leave from kindness reluctantly. Of our four platforms for doing good, it is the simplest, easiest, and most immediately rewarding.

Fairness

Fairness consists of impartial and just treatment or behavior without favoritism or discrimination. As such it is in direct contrast to kindness, which tends both to be partial and to show favoritism, albeit for the best of motives. Fairness seeks to put all participants on an equal footing, regardless of whether any of them has a more intimate relationship with us. As a social mechanism, it is intuitive and informal, the artifact of language and culture rather than law or legislation. Laws may seek to enforce fairness on a broad scale, but in so doing, they will be invoking analytics and theory, thereby taking us into the realm of justice.

Fairness is an amazing ethical accomplishment, one that no other species has ever achieved. To play fair with others entails not just values, the stair that enables kindness, but the additional stairsteps of culture and language, the ones that underpin our ability to reason. Values create the *disposition* to treat others fairly; culture and language provide the mechanism for working out exactly what that entails. Thus, while primates can demonstrate kindness within their own social group and can also detect *unfairness* (especially when it applies to them), they cannot enforce fairness. Only humans learn to play fair, and astonishingly they begin to do so at a remarkably early age.

Standards of fairness vary widely across cultures, but within any given culture, they hold constant. Thus, most children develop a common understanding of what is fair in their culture, even if they don't always abide by it. They are able to do so because fairness is a system based on rules, similar to the rules of a game, something young children pay considerable attention to mastering. One of the fundamental principles of game play is to make sure everyone is playing by the rules. In the case of fairness, everyone must be treated equally—that's the key to the game. Children become outraged when someone does not play fair because, as small citizens confronted by a big world, fairness is one of their best protections.

From a more adult perspective, fairness is best understood as aligning our behavior with the metaphysical force of karma. *Karma* is not a word I grew up with, nor am I deeply familiar with the Eastern tradition from which it comes, so bear with me here. Stripped of its mystical associations, I take karma to be the principle that actions have consequences, that for every action there is a reaction, that we shall reap what we sow. When the Golden Rule says "Do unto others as you would have them do unto you," from a karmic point of view, that means sow good deeds so that you may reap good deeds in return. The universe is in this sense both reflective and reciprocal. It returns to us what we send out to it.

Now, as anyone who has lived in the world can testify, this is far from a perfect truth. Life is not always fair—indeed, sometimes it is spectacularly unfair. Some religions invoke the idea of past lives as a mechanism for restoring the balance, but that is beyond the scope of ethics and is not a metaphysics I care to endorse. Rather, my point is that in the world of social networks, where chance plays a real role, there is nonetheless an abiding principle of "tit for tat" that structures the bulk of the interactions therein. Receiving kindness for kindness, hostility for hostility, and so forth, is not a guarantee. Rather, it is a statistical probability.

Doing good by being fair is an effective and efficient strategy for maximizing the returns from a reciprocal relationship. It is anchored in the concept of *equity*, in which all individuals are treated impartially. As such it asks us to objectify ourselves, to treat ourselves as simply another member of a given set, to be treated equally with all the other members of that set. It asks that we not only hold our own behavior to the standards of equity but that we hold third parties accountable to it as well, even when we are not personally involved. Intervening in such situations can be personally challenging to be sure. The reward, however, is a social safety net that lets us adjudicate potential conflicts to acceptable closure, even with relative strangers.

One final point to note: fairness does not always sit easily with kindness, nor vice versa. The fair thing to do is share things equitably. The kind thing to do is give a disproportionate share to the person most in need. Kindness, in other words, is grounded in *personal* relationships, whereas fairness is grounded in *impersonal* ones. The whole point of fairness is to focus on the *principle* involved rather than the *people*, to create social security in an algorithm that does not depend on relationship capital but rather scrupulously abides by the rules of the game.

Of course, nothing prevents one from being kind after the fact. Once sharing has been done equitably, you can still give part or all of your share to someone else. But you can't do it as a substitute for fair distribution. That jeopardizes the integrity of fairness itself, tearing into whatever fabric of trust it has been able to weave in what otherwise would be a low-trust situation. Indeed, when the conflict between kindness and fairness is writ large, it tends to pit social justice against legal justice, creating serious problems for liberal democracies around the globe.

"Writ large" is the transition we are now going to make. Kindness and fairness are interpersonal in scope. Morality and justice both address a much larger, societal frame of reference. As such, they are more analytical, more structured, and more debatable, for they must cope with a highly heterogeneous set of circumstances while still maintaining a unity of purpose and method.

Morality

Morality consists of principles of conduct, based on shared values and ideas of right and wrong, to which adherents hold themselves accountable. Its purpose is to support trustworthy behavior irrespective of the immediate social context. To operate in such a broad fashion, it must engage the very core of identity itself, calling forth actions that define who we are and what we stand for, regardless of circumstance.

Morality is transmitted through narrative and refined through *hermeneutics*, a form of analytics focused on religious texts and comparable works of art and philosophy. As a specialized discipline, hermeneutics can be pretty abstract, but if you stay close to the narrative, you can usually get the full impact. Take the Old Testament story of Abraham and Isaac, for example. God commands Abraham to sacrifice his beloved son Isaac, to bind him, lay him on an altar, slay him with his knife, and then burn his body. Abraham obeys, but just as he picks up the knife, an angel intervenes, praises Abraham for his faithfulness to God, and points to a ram that is to be sacrificed instead, and thus Isaac is spared. This narrative is core to both the Jewish and Christian tradition. Needless to say, it has given rise to mountains of commentary (here is where the hermeneutics come in), with many conflicting interpretations. But just react to the story, starting with the realization that Abraham's actions are neither kind, nor fair, nor just. Can they really be moral? Where is the benefit to others?

Clearly, Abraham is not acting *directly* for the benefit of others, most obviously not for Isaac, although Abraham's wife, Sarah, could scarcely have been thrilled either. Rather he is fulfilling his commitment to obey God's commands, prioritizing this above even his most intimate interpersonal relationships. This radical departure from social norms positions morality as the highest obligation in life, linking it directly to obedience to a supreme being. Abraham's purpose is to secure and maintain his relationship to God, and he is counting on God to direct his actions to the best outcomes. In the end, his moral behavior ultimately does align with acting for the benefit of others, but only after it has been routed through an unkind, unfair, and unjust trial. That is the whole point of the story.

Historically, religion has been a morally unifying force that, despite its darker issues, has supported networks of voluntary collaboration to deliver social benefits at scale. Anchoring morality in religious doctrine, however, creates a fundamental conundrum for contemporary Western culture. Religions are constructed around sacred narratives. Such narratives must remain constant even as the secular narratives around them change. The result is a growing divergence between static religious narratives that are to be conserved and dynamic secular narratives that must evolve. In contemporary society, this has led to a schism between two highly polarized worldviews, one anchored in evolution, the other in creationism. Evolutionists, by and large, feel intellectually superior to creationists, while creationists feel morally superior to evolutionists. This is not a debate worth joining, but it is definitely a schism worth healing.

A foundational function of religion is to confer upon morality a transcendental authority. Moral tenets derived from hermeneutic analytics and applied to sacred narratives culminate in a code of ethics that has divine authorization. Moral actions reinforce one's relationship with the divine, and immoral actions jeopardize it. When the religious narrative incorporates life after death, an eternity of consequences, then the authority of morality in this world becomes absolutely paramount. This is the platform upon which Western ethics developed over the past two millennia.

Now take all that away. Replace creationism with evolution as society's primary existential narrative. Religious narratives can still be valued, but they will no longer be treated as sacred. If and how one connects to spiritual support is a personal choice, and life after death is just another narrative, and not a particularly credible one. There is nothing in the Darwinian narrative that can confer upon morality an absolute status. It has been cut loose from all its traditional moorings. Whatever authority it has must be derived from social consensus. In this context, morals look a lot more like *mores*. They transcend the personal, but only to the boundaries of the social group. Most importantly, they have lost any a priori claim to being preeminent over all other commitments. Right and wrong are still important, but they have become relative, not absolute.

The impact of this shift on our traditional ethics has been slow to emerge. For several centuries Western culture, leveraging a received social consensus, has been able to benefit from the inertial momentum of Judeo-Christian morality without having to commit to Judeo-Christian metaphysics. But that societal consensus is now unraveling under the pressures of secular education, economic dislocation, technology disruption, gender diversification, political polarization, abusive rhetoric, fake news, and a host of other contemporary forces. As a result, we are losing the common ground upon which to base tacit support for traditional ethics. Public discourse has become less civil and more cynical, repositioning virtue as either naive or self-serving. With no external source of truth, we are thrown back upon our own devices. Now what?

We know we must engage one way or another, so what we are looking for initially is simply a reasonable place to start. A good first step is simply to conform to the ethics of the social group in which we find ourselves at present. These may not be the ethics we really want, but conforming to them buys some much-needed time to reflect, and it keeps us from getting tangled up in conflicts we will likely have little stake in. In parallel, we can keep a lookout for role models whose words and actions touch us deeply. These will be people to whom we are instinctively drawn, whose memes attract us and resonate with our experience to date, whose behavior we feel inspired to imitate.

Embracing that imitation is a good next step, even if at first we do not fully comprehend the underlying moral vision that is guiding our actions. Competence, in other words, can precede comprehension. As in academics or athletics, we normally learn from imitation first, then later through understanding the underlying principles. Revered texts help mediate between these two states, narrative providing the grounds for imitation, hermeneutics, the foundation for understanding. In a secular worldview, however, such texts are never sacred. They are only an enabling aid, often works of literature rather than of religion or philosophy.

The next step in our progression is to declare and embrace our own moral vision, organized around the highest metaphysical force we can imagine, something we want to live for, and, should the necessity arise, something we are willing to die for. This could be family, or love, or social justice, or artistic creation, or any other cause that deeply enlists our heart and mind. Whatever it is, it will anchor our strategy for living, becoming the still point in our turning world, the core in relation to which everything else is context.

We can secure this commitment by embedding it in a narrative that incorporates the role models we want to emulate, the memes we want to embrace, and the behaviors we want to support or oppose. In this way, the ethics in which we have become competent are now inscribed within a narrative that is inspirational, sensible, and consistent with whatever other narratives we have embraced. By combining this metaphysical narrative with our commitment to an ethical code, we become philosophically complete and morally coherent.

What separates narrative-based ethics from traditional religion is the absence of any a priori claim to truth, any grounds for unilaterally imposing moral obligations onto other people. We have witnessed such claims play out over several thousand years. Despite the best of intentions, they have consistently been abused by those in power such that our holiest sentiments have repeatedly become platforms for exploitation and genocide. It is no accident that the secular worldviews of the Enlightenment were developed on the heels of a century or more of religious persecution. In an early draft of the Declaration of Independence, Thomas Jefferson wrote, "we hold these truths to be sacred and undeniable." In the final draft, *sacred and undeniable* was changed to *self-evident*. For a secular state to guarantee religious freedom, it must forsake transcendental authority.

Inside a community of faith, moral authority can be absolute, but across communities of faith, it cannot. To operate across independent communities requires a different kind of social network, one that is secured by legislation rather than faith. For that, we must turn to our fourth and final domain of ethics: justice.

Justice

Justice has two meanings that intersect with our exploration of ethics. The first is the use of power as appointed by law to support fair treatment and due rewards. This we will term *legal justice*. The second is the quality of being impartial, fair, decent, and morally right. This we will term *social justice*. It is essential to keep these two concepts separate because, in a secular state, they derive their authority from very different sources.

The goal of legal justice is to use the laws and institutions of the state to apply the principles of fairness to the governance of society as a whole. In a secular state, there is no transcendental narrative to authorize this effort, so the edifice of justice is instead built upon theory. The secular theory of legal justice is rooted in the concept of a *social contract*, an implicit set of norms that establishes the nature and boundaries of fairness for any given society. This is the foundation upon which legal justice is constructed.

Getting agreement as to what this social contract entails, however, is challenging and gives rise to political parties with competing points of view. In a well-regulated democracy, such competition is a good thing as it allows society as a whole to adapt its social contract over time to changing social conditions. But in order to have a well-regulated democracy, there has to be a foundational agreement with respect to first principles. Lacking a transcendental narrative, how can we get to this underlying agreement?

A good place to begin is with a thought experiment proposed by John Rawls in his 1971 book, *A Theory of Justice*. Rawls was seeking a mechanism by which a truly fair social contract might be brought into being, even hypothetically. He asks us to imagine a committee of rational, self-interested individuals charged with creating a social contract "under a veil of ignorance." What that veil blocks from view is what role anyone on the committee will actually play in the envisioned society. That is, one might end up being rich or poor, male or female, young or old, healthy or infirm, of one race or ethnicity or another. It behooves everyone on the committee, therefore, to protect the interests of every role as best as possible. This mechanism, Rawls asserts, will result in a just social contract, provided the committee itself is sufficiently diverse to represent society as a whole.

In actual fact, however, social contracts are normally determined by an elite, the United States being as good an example as any. In that context, I think it is truly commendable that our founding fathers (all male, all white, most well-to-do) were able to draft a constitution as fundamentally just as ours is, especially since they included a mechanism for amending it, thereby allowing, among other things, for future inclusion of initially excluded constituencies. A testimony to the enduring power of that constitution is that today most US citizens enjoy the privilege of living under rule of law.

Rule of law is the foundation of Western democracy. As defined by the World Justice Project, it consists of four core principles: accountability, just laws, open government, and accessible and impartial dispute resolution. Under rule of law, individuals cannot break the law with impunity, legislatures cannot pass laws that violate the nation's constitution, elections and voting must be public and auditable, and litigation and mediation services must be available to all.

Rule of law is essential to securing individual freedom. This is what makes America, despite all its shortcomings, so attractive to immigrants. We live in a society where you do not have to bribe someone to get your due, and people get to keep what they earn (after paying taxes). The same is not true for the majority of people living elsewhere today. Whether it be from external threats like terrorism, ethnic genocide, or military attack, or internal threats like corrupt officials, autocratic dictators, or organized crime, well over half the world lives without reliable access to legal justice. However flawed our system of justice may be, we should not underestimate how valuable a gift it is, nor how precarious it is to maintain.

For legal justice to hold sway, the vast majority of citizens must voluntarily comply with the law, even when it works to their personal disadvantage. This is an ethical obligation. We must not only acknowledge the social contract; we must commit to its essential principle of equitable fairness and defend it. We cannot outsource legal justice to public agencies and institutions and be done. There are not enough courts or police to impose lawful behavior from without. If we are to enjoy rule of law, support for it must come from within our own communities and be integrated into our everyday lives.

I emphasize this point because legal justice by itself is quite an accomplishment and should never be taken for granted. Indeed, there are many who argue we should not ask more of government than simply to deliver this. To stop here, however, leaves out the question of *social justice*. It is not illegal for people to be homeless, to lack proper health care, to be poor, to be uneducated, to be disadvantaged in any of so many other ways. But it is inhumane to allow people to suffer such conditions when the means to alleviate them are available. The question is, Is it *socially unjust* to do so? And to the degree that it is, how much should government intervene to remediate the situation?

In modern Western democracies, the general answer to these two questions has been *yes* and *some*. That is, the social contract has been extended to provide safety nets for health and welfare, to promote equal opportunity for the disadvantaged, and in some cases to make reparation for prior acts of discrimination, including slavery and genocide. Where legislated, such efforts are funded by taxation intended to redistribute wealth and resources, the goal being to level the playing field and create a more just and stable society. Reasonable as all this may be, however, it puts considerable stress on the institutions of society, in some instances generating a backlash that puts the entire social contract at risk. If we are going to address this situation, we need to manage this stress thoughtfully.

To begin with, we need to determine how level a playing field we want to create. All societies of any size entail a persistent inequality of wealth, often to an extraordinary extent, so what is the socially just thing to do about it? Once again Rawls has proposed an answer. Inequalities of any kind, he argues, can be considered just if they work to the benefit of the least advantaged group. Thus, for example, if higher incomes are taxed disproportionately to pay for social services, the least advantaged are benefiting from that extra wealth, and therefore it is just that it exists. This principle has been adopted by most Western democracies and as such has taken on something like the mantle of a universal social contract.

Unfortunately, Rawls's idea works better in principle than in practice. To begin with, while the intent is to pay for these services by disproportionately taxing the rich, it turns out the rich are very good at protecting their assets and income. They can afford to hire legal and financial advisers to help them minimize their tax payments, aided in many cases by legal loopholes inserted by politicians who have been funded by these very same folks. In any event, they rarely end up paying anything close to their fair share. The actual burden of taxation is borne further down the economic ladder, falling on citizens less well positioned to bear it. This result is categorically unfair and creates widespread resentment against the uber-rich as well as cynicism about the system overall. Meanwhile, at the other end of the spectrum, while public services, following Rawls's principle, do indeed improve the welfare of the least advantaged, they often leave the group just above them, what we sometimes call the lower middle class, underserved. This also is unfair, and it too creates a breeding ground of resentment, contributing to the rise of racism, separatism, and xenophobia we are presently seeing around the world.

The problem here is not so much lack of empathy as it is limited resources. It is exacerbated in several ways:

- Publicly funded services are deployed with bureaucratic controls in order to prevent fraud and abuse. Such bureaucracies entail a myriad of jobs that are expensive to maintain, and they end up consuming a significant portion of the resources they are supposed to be redistributing. This creates a crisis of *efficiency*.
- The demand for publicly funded services vastly exceeds the supply, but everyone who qualifies is entitled to them regardless. This inevitably leads to unmanageable workloads and extended delays in getting promised services. Despite good intentions, the system itself becomes a wearisome burden for everyone involved. This creates a crisis of *effectiveness*.
- Finally, any set of benefits targeting social justice, once put in place, becomes an ongoing entitlement. As such, it must be funded in bad times as well as good. Generous gestures in times of plenty can lead to fiscal emergencies in times of want. This creates a crisis of *financial viability*.

In sum, modern Western democracy's efforts to support social justice through legislated programs have had mixed results. Publicly funded programs and systems rarely live up to the high expectations political rhetoric sets for them. Often, they fall well short, in the worst cases giving rise to what is now a growing number of failed states. For once it becomes clear that the social safety net is failing, a disaffected populace rejects the establishment version of the social contract and opts instead for the more accessible patronage of a local leader. Typically, this leader operates outside the law, often abusively so. From an outsider's perspective, such an alternative seems radically counterproductive, but to the already disenfranchised, it can look like the best available bet. *The state will never take care of us,* they reason, *so we need to align with the powers that can, whatever that may entail.* The result is a society that has abandoned not only its institutions for social justice but those for legal justice as well.

Failed states have emerged in Latin America, Africa, Asia, and the Middle East. In every case they result in horrific living conditions, driving massive flights of refugees, exporting their crises to the entire world. To help meet this threat, Western democracies must find new ways to improve the delivery of public services that support social justice. Strategies for so doing include addressing:

- *The crisis of efficiency,* in part by replacing centralized people-intensive bureaucratic controls with decentralized technology-intensive transaction systems enabled by cloud computing, mobile communications, artificial intelligence, blockchain record keeping, and the like.
- *The crisis of effectiveness,* in part by increasing productivity with real-time, personalized information systems similarly enabled by advanced technologies.
- *The crisis of financial viability,* in part by engaging corporations to dramatically increase their voluntary support for private philanthropy and nongovernmental organizations.

At the same time, we must acknowledge that social justice has become a highly politicized issue that is creating deep divisions in Western society. Current programs and systems for delivering social justice struggle to keep up, and it is not clear there is the political will to remedy the situation. As we have just noted, however, failure to support social justice puts legal justice at risk. To enjoy rule of law, one must accept duty of care as an ethical obligation.

Concluding Remarks:
How well can we expect to do good?

Our analysis of doing good has sought to align traditional ethics with contemporary metaphysics. In so doing, four dimensions of ethics—kindness, fairness, morality, and justice—are positioned at four different places along our metaphysical staircase. One inference to take away from this is that the longer a given dimension has been tested by time, the more likely it is fit for purpose.

In that context, kindness is where we can expect the highest probability of success. If all you took away from your engagement with ethics was *Be kind,* you would not be too far off the mark. True, sometimes kindness can be camouflage for conflict avoidance. When we do not intervene in a situation that calls for criticizing someone's behavior, when we "kindly" let it go instead, often it is because we are fearful of the consequences of engaging. This is not being ethical; it is being cowardly. That said, kindness is still our best overall bet, and the more of it we can show, the better.

Fairness comes next. It has a broader scope—we can be fair to people we would not necessarily want to be kind to—but it is also more complicated to enact. Often it can require dialogue to reach a common understanding of what is the fair thing to do, and such conversations can be challenging to manage. As an aid, we can put Rawls's veil of ignorance to good use. That is, if you did not know which side you were on, if you could end up on either end of the decision, would you endorse the outcome regardless? Most of us have some ability to imagine ourselves in another's shoes, so this tactic can be of real help. A more vexing question is, Do we have the ethical courage to stand up for an unpopular conclusion? For fairness to work, we have to be brave, and sometimes we fall short of that mark. Nonetheless, most of us can aspire to be fair in our own dealings with people with whom we do not have a high-trust relationship, and this does provide a reasonable foundation for social justice on a local basis.

The next step up in the staircase takes us to morality, the home of ethics as it is traditionally conceived. In the context of a secular worldview, the critical question we need to answer is, What authority can morality have if there is no God? We saw that it could have social authority within a given community, but that it could not claim transcendental authority over other communities. Social authority is fine as far as it goes, but it does not inspire or engage the personality to forge its own moral identity. For that we turned first to inspiring role models and then to narrative, noting that every culture circulates stories that celebrate ethical personalities, and that such stories do indeed engage the personality to develop its own moral identity. That said, we left an awful lot of open space when it comes to determining which narratives are sufficiently worthy to embrace. Secular morality in this guise is at best a few centuries old, and we are still inventing it as we go along. We know we are going to make mistakes. We need to stay humble even as we seek to stay true.

Finally, we come to justice, and here too we are engaged in a work in progress. While legal justice has a few thousand years of time-testing under its belt, state-funded investments in social justice at scale are less than two centuries old. Few deny they are fundamental to maintaining the stability and integrity of the modern state, but there are also powerful forces working to undermine this stability. We are a long way from solving the crises of efficiency, effectiveness, and financial viability. As humans we have proved we can rise to the occasion if need be, but we have also shown a tendency, in the absence of a crisis, for our attention to wander—hence the political wisecrack, "Never waste a good crisis." It seems likely we will have plenty of opportunities in our future to put that dictum to the test.

So, in the end, note to self—when it comes to doing good, here are four key takeaways:

1. Be kind. *Check.*
2. Play fair. *And make sure others do, too.*
3. Act morally. *At minimum, imitate virtuous people to improve your moral competence. Eventually, integrate virtue in your life narrative to achieve moral coherence.*
4. Advocate for justice. *This is not a sprint, nor even a marathon. It is a relay race that will carry on long past your time. Run your leg as best you can.*

CHAPTER 10

Being Mortal

This chapter is the last in our contemporary take on metaphysics and ethics. Mortality plays the same role in a secular worldview that immortality does in a religious one. It is the ultimate ground upon which we play out our lives, the destiny that shapes our ends. It is here that ethics and metaphysics must meet and shake hands if we are ever to be fully integrated.

Being Mortal is the title of a book by Atul Gawande, one I will be drawing upon shortly. What I like about the phrase is that, once again, we can take advantage of an ambiguity in the word *being*. Earlier, we invoked it as a noun, contrasting *being* with *Being*. Here we are using it as a verb, contrasting two different senses, one expressing state, the other expressing action. If we take *being* as expressing state, as in the phrase "being tired," then *being mortal* means experiencing the condition of mortality and making one's peace with it. If we take it as expressing action, as in "being creative," then *being mortal* implies a way of acting in the world, a predisposition to prefer certain values and pursue certain ends. Both senses deserve our attention.

With respect to how we experience the condition of our mortality, there has been a tacit consensus for more than a century that loss of religious faith has left us fearful in the face of death. The conventional view is that a divine creator overseeing an eternal afterlife gives us the most desirable answers to our anxieties about death. Thanks, however, to a host of secular trends, that narrative is no longer as credible or reassuring as it once was. This raises the question of whether there is a secular narrative that is both credible and capable of addressing our needs.

The Darwinian narrative we have been developing throughout this book is certainly credible, but can it address our anxieties about death? Here, in synopsis form, is what it has to say:

1. As human beings, we are a product of evolution.
2. Evolution, in turn, is a product of natural selection.
3. Natural selection, in turn, is a product of life-or-death competitions.
4. If there were no death, there would be no natural selection.
5. If there were no natural selection, there would be no evolution.
6. If there were no evolution, there would be no us.

Without death, you and I and everyone we love, and every living thing we see, and every artifact we admire, simply would not exist. Strange as it may seem, it appears we all owe death a big show of thanks. At minimum, it is hard to see how we can cling to a view that death is unfair or unjust. Death is how we got here. It is an integral and necessary part of our presence. We can't repudiate it. We have to come to terms.

Of course, none of this deals with our anxieties about what might happen to *us* after death. Nor does it deal with our fear of death itself, something that has also been built into life via natural selection. As descendants of those who have survived, we are predisposed to fear and flee death. That too is an integral and necessary part of our presence. To make things even more challenging, our heritage of language, narrative, and analytics enables us to bring these fears to mind even when we are perfectly safe. Unlike our mammalian cousins, we can *worry about* dying, be that in reference to ourselves or to anyone we love. What will happen to them? What will happen to us? What will happen to me?

This, of course, is our ego speaking, and as we know, it can be somewhat fragile. To function effectively in the world, we must find some way to assuage its anxieties. At the dawn of human history, we met this need through pure narrative, creating myths that explained the otherwise inexplicable. Later, with the addition of analytics, we developed religion, which has served us admirably over the centuries, although not without showing its darker side as well. Through religion, death is contained within a larger system of eternal value, freeing us to get on with our daily lives. Now, however, our present age has cast a secular shadow over this narrative. The modern ego senses a chill in the air. It is haunted by a refrain from medieval liturgy: *Timor mortis conturbat me*—The fear of death disturbs me.

We first experience this fear when losing someone we love to death. Grief runs deeper than we are able to acknowledge and we are in great need of comfort, but the secular shadow cuts off access to the light and warmth we crave. Living together with loved ones is at the core of our being. We are mammals first and foremost, family members, tribal by nature. We need to keep whatever makes us part of that bigger community alive and present within us. We need our parents and grandparents, our missing friends and lovers, our memories and traditions, present in our lives. It is hard to be whole without them.

The task of restoring that wholeness falls first to narrative. Whatever we think we may be doing in life, what we are actually doing is telling stories and listening to others do the same. We tell stories about ourselves, our communities, our plans, our work, our failures and successes, our loved ones, past and present. That is how we maintain both identity and continuity. Without these stories, we are no one. So the first requirement we can place upon any narrative that embraces mortality is that it must help us maintain our identity and continuity, help keep us whole and together.

That means it has to be both *about us* and *not about us*. It must be about us, for we are the ones who need to be served. But it cannot be about us alone because we know we are temporary and transient. As actors we come and go, but our play can go on when we are no longer here, just as it has been going on long before we ever got here. Most importantly, whatever this play is about, it must confer meaning on the limited time we have here together.

Our culture has served up many such narratives over the years. A favorite of mine is a poem by Robert Frost, one of the best meditations on mortality I know of:

After Apple-Picking

My long two-pointed ladder's sticking through a tree
Toward heaven still.
And there's a barrel that I didn't fill
Beside it, and there may be two or three
Apples I didn't pick upon some bough.
But I am done with apple-picking now.
Essence of winter sleep is on the night,
The scent of apples; I am drowsing off.
I cannot shake the shimmer from my sight
I got from looking through a pane of glass
I skimmed this morning from the water-trough,
And held against the world of hoary grass.
It melted, and I let it fall and break.
But I was well
Upon my way to sleep before it fell,
And I could tell
What form my dreaming was about to take.
Magnified apples appear and reappear,
Stem end and blossom end,
And every fleck of russet showing clear.
My instep arch not only keeps the ache,
It keeps the pressure of a ladder-round.
And I keep hearing from the cellar-bin
That rumbling sound
Of load on load of apples coming in.
For I have had too much
Of apple-picking; I am overtired
Of the great harvest I myself desired.
There were ten thousand thousand fruit to touch,
Cherish in hand, lift down, and not let fall,

For all
That struck the earth,
No matter if not bruised, or spiked with stubble,
Went surely to the cider-apple heap
As of no worth.
One can see what will trouble
This sleep of mine, whatever sleep it is.
Were he not gone,
The woodchuck could say whether it's like his
Long sleep, as I describe its coming on,
Or just some human sleep.

The metaphor of apple-picking describes a life's work gifted with an abundance of opportunities and the good fortune to have done well by them. From the perspective of mortality, it doesn't matter what that work actually is, whether it is conducted in the home or in the marketplace, only that it contains elements, like the apples, that one values deeply enough to cherish and not let fall. Such work, Frost suggests, is not meant to last forever. At some point, we become overtired with the great harvest we ourselves desired. What intervenes to cause this change of heart is a new perspective. The world, seen through the lens of aging, has begun to blur, the mind to waver as when it falls asleep. This is death encroaching, but it is not a frightening force. Like the woodchuck's hibernation, like sleep itself, it is a totally natural phenomenon. Every living thing that has ever been has died. How bad can that really be? To paraphrase the last line from another Frost poem, "One could do worse than be a picker of apples."

Such is the way in which a secular worldview makes its peace with death. We contemplate mortality in times of reflection to find a narrative that we can live with. That sets us up for the real work at hand: not contemplating mortality but actively *being mortal*, the work of everyday life.

Let me suggest at the outset that being mortal is not the work of a lifetime. First, there is a time to be immortal—as a child, as an adolescent, perhaps well into one's twenties. Yes, we must embrace the importance of safety, both for ourselves and others, but taking on personal mortality at this age is more overhead than it is worth. Better to let the horses run, to pursue playing, education, friendships, love, a job, a career, all with an infinite horizon—to, as the song says, "take your passion and make it happen."

To be sure, somewhere in our late twenties or early thirties we will have to come to terms with the contradictions and limitations in our life history to date. Even here, however, I think we can defer being mortal a while longer. This is a time to take responsibility, as best we can, for our own karma, to work on our faults, clean up our messes, embrace the things we love, and, in short, become adult. To add being mortal to this mix is, in my view, too much to ask. Just getting through intact would be good, and if we can accomplish some key goals in the process, so much the better.

Assuming no crises of health or welfare, it is perhaps somewhere between the ages of 50 and 60 that being mortal starts to become an interesting option. We realize now that the game we are playing has entered the second half. There is still plenty of time on the clock, but there is no longer an infinite horizon. This is a time to get serious, to take stock of who we are and where we are going, to double down on the things we care about most, to begin to extract ourselves from the things that are holding us back. We are now, in all likelihood, as capable as we have ever been or are ever going to be, so our focus should be on how to put our power in service to whatever we honor most. Embracing being mortal at this stage simply means we are no longer willing to waste large swaths of our time.

Let us call this *being mortal, part 1,* something we will delve into more deeply in a moment. But before we do, let's look ahead to what we might call *being mortal, part 2.* This is the last major phase of life, and it typically begins with a breakdown in some core part of our body, an illness or injury from which we may recover, but not fully. Instead we experience for the first time a genuinely irrecoverable decay in our capability, one that makes us viscerally aware we are now well into the fourth quarter of the game, and the clock is ticking. This is the phase of life that Gawande, a physician, addresses so clearly and valuably in his book. Part of his approach is to stretch the traditional idea of hospice beyond simply providing palliative care at the very end of life. Instead he asks us to reach further back in time, to incorporate strategies and tactics that can maximize quality of life over a considerable period while still coping with diminishing capabilities.

The main difference between these two stages of being mortal is that in the first, we are potentially at the height of our powers, and in the second, we are fighting a rear-guard action against losing them. Both are part of being mortal. I am going to focus the rest of this chapter on part 1, but I am going to steal a tool from part 2 to help frame the discussion. That tool is a set of five questions that Gawande returns to several times during his book as a kind of checklist to help hospice patients address and improve the quality of the life remaining to them. Here they are:

1. *Do you understand what is happening to you, and what your options are?*
2. *What are your biggest hopes and fears?*
3. *What are your goals?*
4. *What trade-offs are you willing or unwilling to make?*
5. *Is the course of action you are pursuing consistent with your answers to the prior questions?*

These are incredibly important questions to discuss with anyone approaching end of life. They are not the sort of thing we normally talk to each other about, and in some cases we may need to have someone facilitate the conversation. Once the answers to these questions are in plain view, however, a huge burden is lifted. Everyone can now focus on making the very best of the time left, without second-guessing decisions that otherwise would be very hard to make.

Let me propose we should pose these same questions to ourselves at the very outset of being mortal, part 1. This is a time, I am hoping, when we are close to the height of our powers. How might we answer these questions in that context? What difference would it make to our ethical decision-making at this time in our lives?

Do you understand what is happening to you, and what your options are?

Arguably, the whole point of this book has been to answer this one question. So let me recap what it has said. We are born into this life conscious, with a nervous system that communicates pleasure and pain, and a hormonal system that activates desire and fear. Through nurturing we absorb maternal values and through parental discipline, paternal values. Well before we learn to speak language, we are learning cultural practices for eating, dressing, playing, and hygiene. From the very start, we are hearing language, and as we become proficient in it, we adopt it for instrumental purposes to pursue our desires and protect against our fears. In parallel we hear and participate in all kinds of narratives, from family conversations to storybooks, media, and the like.

By the age of four we are building our own narrative about ourselves, one we will carry forward continuously for the rest of our lives. From that point on, narrative does most of the heavy lifting. We understand the world—ourselves, our loved ones, our plans, hopes, fears, desires, the intentions of others—all through the mechanism of narrative. We are constantly telling stories to ourselves and to each other. Over time our very own selves become the sum of our stories. Yes, there is a material world, and objective events unfold there without the need of narrative—births and deaths, the accumulation of wealth or debt, physical changes to us and to others, environmental changes in our community and society. But we *understand* these changes, we internalize them, via narratives, refined over time through analytics. Narrative is our primary imagination, analytics our secondary imagination. The latter helps us edit the former, the former drives our choices in life. And we continue in this manner until we die.

That, in essence, is what is happening to us. We are acting out our lives as characters in an interlaced series of stories. Ethics shape the roles we have chosen to play and how we play them. They align us with the narrative, and they keep us in character. Some of our stories are told to us by others; some we tell ourselves. Sometimes we're the windshield;

sometimes we're the bug. Sometimes we are just a third party, perhaps the narrator, perhaps the audience. One way or another, we are part of the story. Story is what holds the whole fabric together.

Now, if this is our situation in life, what implications does it have for being mortal, part 1? What are our options? Essentially, I see three broad paths forward:

1. *Convert mortality into immortality through religious faith.* This, in effect, changes the story radically. It is a perfectly viable option, but it is one we are leaving to others.
2. *Ignore, defer, or repress the fact of our impending personal death.* This is easy to do, for there are always plenty of activities to distract us. We can essentially ignore being mortal, part 1, and carry on instead as we always have.
3. *Embrace personal mortality and reframe our narrative accordingly.* This is the option we are exploring here.

To reframe our narrative around being mortal, part 1, let us imagine ourselves in the position of a professional athlete nearing the end of his or her career. The players around us have gotten younger, and the new ones have talents and energy to burn. We do not. Unlike them, however, we have the advantage of experience. We know the game, we know our strengths, and we can still have a lot of success, provided we play to those strengths and mind our limitations. In that frame of mind, we can tackle the remainder of Gawande's questions.

What are your biggest hopes and fears?

Hopes and fears originate on the stairstep of desire. From there they emerge into consciousness, where they motivate much of what we do. We, in turn, do our best to engage them in accordance with our values. We are not in control of our hopes and fears, but we can learn to manage them if we can acquaint ourselves with them sufficiently. That, it turns out, is easier said than done, hence the importance of this question.

In the context of being mortal, part 1, identifying our biggest hopes is part of prioritizing where we allocate our no-longer-seemingly-infinite resources. Time has become precious. Where do we want to spend it?

Most of our lives we typically do not prioritize our biggest hopes, in part because they are so big. That makes them both hard to socialize and challenging to execute. But we are entering the now-or-never part of the game, and after all, these are our biggest hopes, so why not? What have we got to lose?

Of course, that brings in the other half of the equation: What are our biggest fears? Some of these go way back to childhood—nightmares that never completely came out from under the bed. Others are newly arrived, some carrying the chill of our mortality with them. Unlike hopes, which can become more powerful when exposed to conscious analytics, fears normally become less powerful. It behooves us, in other words, to bring them into the light.

Taken together, hopes and fears are what the Enlightenment philosophers referred to as *the passions*. As David Hume famously said, "Reason is, and ought only to be, the slave of the passions, and can never pretend to any other office than to serve and obey them." The passions are the energy and power that drive human action and change. The whole point of addressing our hopes and fears is to put ourselves in touch with their force. Once we have done so, reason—what we have been calling the conscious ego—can direct that energy to achieve our goals. That becomes the focus of the next question.

What are your goals?

Goals are one of the great "good tricks" of human culture. They transform our life's narratives into plots. They give our actions structure and meaning. They help us track our progress toward the achievements we seek. They make us more effective and efficient. That's why they are so popular. Like all good tricks, however, goals themselves are amoral. They can be readily co-opted for trivial ends, as any number of digital apps make clear. They can also be put to immoral, or even horrific, use. So when it comes to our own goals, it is important we choose wisely.

For our present purposes, we have confined our choosing to the zone of being mortal, part 1. Prior to this time, when we are in effect "being immortal," goals are used broadly across a wide variety of circumstances to allocate our time and focus our actions on the results we seek. At the other end of the spectrum, during being mortal, part 2, goals are typically narrowly defined targets to help us manage our way through day-to-day circumstances that are drifting out of our control. With being mortal, part 1, by contrast, the goals can be ambitious, provided they are focused and prioritized.

If we keep in mind our imagined status as an aging pro athlete, the two things most worth prioritizing at this stage are *legacy* and *impact*. Legacy is our contribution to the ongoing narrative, the one that preceded our arrival on the scene, the one that will carry on after we leave. We are indelibly part of that story—but can we be, and do we want to be, a more memorable part of it? If so, what do we want to be remembered for, and what can we do now to reinforce that contribution? Ironically, having less overall time at our discretion, we actually have more freedom to allocate that time asymmetrically. That's where the idea of impact comes in.

Impact is a function of what one might call "fish to pond ratio." Any fish can be big or small, depending on the size of the pond it swims in. To have impact, we need to be a big fish. So the search is not for the biggest pond, but rather, the biggest pond in which we can make a material impact. Think, for example, of monetary donations. A thousand dollars can be life changing in one context and a drop in the bucket in another. Same goes for 20 dollars, same for one million.

More importantly, the same goes for our time. An hour a week doesn't seem all that impactful in the context of a year, but how about if those 50 hours were spent just with one person or devoted to just one cause? And what if they were concentrated over the course of, say, one month? It's all a matter of asymmetry. The more asymmetrical our allocations, the greater the impact we can have. Power is not, in other words, reserved for the elite few. We can all be powerful.

We should note, however, that asymmetrical actions normally generate resistance. That is, when we step outside our norms, that can be upsetting to people who have very different expectations of us, expectations we ourselves helped to create. To serve our goals, we are going to have to do some heavy lifting to reset those expectations. That is what leads us to our next question.

What trade-offs are you willing or unwilling to make?

To begin with, we need to acknowledge that there is another force competing for our attention during being mortal, part 1: the "bucket list." It represents a host of deferred pleasures and experiences that we have hopes to enjoy before we leave the planet. These hopes have their own claims to make on our time and resources. Which ones are we willing to trade off to make more progress on legacy and impact, and which not?

Of course, here we may be dealing with more of an *and* than an *or*. One of the great advantages of being mortal, part 1, is having more discretionary time than we are used to, so there is reason to believe we can have at least some of our cake and eat it too. But there is another, more subtle force making claims on us as well: our routines. We've spent our whole lives spinning up a gyroscope of routines to keep our days in balance. Now we are looking to tilt it in a new direction. We should not be surprised at its power to keep our discretionary activities confined to well-trodden paths.

There is no need to seek heroic solutions to this kind of challenge. The key point is not to let trade-offs defeat us. We can push back against them, but only to a degree. Instead, employing the fish-to-pond ratio principle, once we have a sense of how big a fish we think we can free ourselves to be, then we can target the kind of pond where we can realistically make a big splash.

Splash is important, and not just because it gratifies the ego. It signals a story worth telling, and retelling. This amplifies the impact of a local act, inspiring others to imitate it in other circumstances. As the "Good News" sections of the Internet keep reminding us, even the smallest gesture can have a radiating effect. That is why any of us can aspire to create legacy.

Aspiration, of course, is not the same as actualization. And that brings us to our final question.

Is the course of action you are pursuing consistent with your answers to the prior questions?

For most of us I think the most common answer might be "sort of" or "maybe" or "in a way." The challenge, as Stephen Covey taught us to see, is that we have spent most of our life prioritizing the *urgent* over the *important*. Now it is time to lean the other way. We need to direct our time and attention to the most important things first. We need to build some new habits. How do we do that?

One thing we have going for us is that we have embraced the fact that we are mortal. This allows us to take being mortal, part 1, seriously—something, frankly, we have never done before and, indeed, many people never do at all. We also have the time and capacity to reflect on our next steps. In so doing, more than anything else, we need to get clear on who and what we want to be in service to. This is a deeply personal decision, and we need to make it as explicitly as we possibly can.

Once that is clear, then we can start building the routines that make that service natural, even inevitable. A simple start is to begin each day by asking ourselves: Is there something we can do today to further our cause? What might that be? How could we go about doing it? Just asking the question will put the important in front of the urgent, and the more we are able to act upon our thoughts, the more we ingrain the new routine.

This is a pleasingly undramatic conclusion. Its claim is that there is no significant barrier preventing any of us from creating a legacy that has impact. Not only can we craft a narrative that supports being mortal, we can act upon that narrative to achieve ends that have meaning and value beyond ourselves.

Conclusion

We have come to the end of our contemporary take on secular metaphysics and ethics. Our goal has been to combine frameworks from the physical sciences, social sciences, and humanities, along with whatever common sense we could muster, to address two fundamental questions at the core of all philosophical reflection.

The first is simply, What is going on? We took the most cosmic view we possibly could, reaching all the way back to the Big Bang to find some principle that could span the 13.8 billion years between then and now. In so doing, we called out the Second Law of Thermodynamics, identifying entropy as that unifying thread. Our interpretation of that law asserts that the universe as a whole, as well as all processes on Earth, continually and perpetually work to increase entropy in every way possible. This, we argued, is what leads to the emergence of complexity. Ever-increasing degrees of order, because they create more disorder than they displace, are the unintended consequence of increasing entropy. By tracing the evolution of that complexity up our infinite staircase, a philosophy of materialism, grounded in the physical sciences and fueled by emergence and self-organization, can account for everything we see and know around us.

What materialism does not do justice to, however, is the *experience* of consciousness and its interactions with everything that lies above it on the staircase. For that we turned to the social sciences, specifically to the concept of *strategies for living* and the forces of natural and sexual selection that underlie the Darwinian theory of evolution. These helped us understand how mammalian socialization has shaped the emergence of values, culture, and, ultimately, the invention of language.

What Darwinism does not do justice to, however, is how radically language separates human beings from all other creatures. We were able to maintain continuity with Darwinism while developing the upper part of the staircase by forging an analogy between *genes* and *memes*, one that showed how cultural forces akin to natural and sexual selection lead to the emergence of narrative, analytics, and theory. These last three topics we discussed primarily in relation to the humanities, thereby including in our survey all three of the major foundations of liberal arts education.

In so doing, we never reached out to incorporate an idealist perspective, being guided instead by a consistent thread of pragmatism. The absence of idealism, however, became troubling when we sought to transition from metaphysics—what is going on?—to the critical questions for ethics—what is goodness, and what are we are supposed to do about it? Doing good requires spiritual support, and nothing in our staircase spoke to that need. Transcendentalism does fill this need, but it entails departing from a secular worldview and embracing an idealist narrative of belief. That put us in a quandary. We staked out a pragmatic middle ground by asserting that spiritual support verifiably emerges at the level of consciousness as witnessed by the practice of mindfulness. We claimed that mindfulness allows anyone to experience being as Being, thereby securing the spiritual support needed for ethical living without requiring any particular belief about the underlying ontology involved.

With that foundation in place, we turned our attention to ethics. In the context of the infinite staircase, our primary goal was to present goodness as an emergent phenomenon that evolves from the bottom up rather than devolves from the top down. We discriminated three types of goodness, each emanating from a different stair—personal pleasure from desire ("feels good"), effective tactics from culture ("works good"), and caring for others from values ("is good")—and we focused our discussion of ethics on the last.

To care for others beyond one's immediate family and friends calls for altruism. Altruism does not emerge until we reach the stairstep of narrative. Narratives, accompanied by analytics, enable altruism through the action of memes that bind to our identities and shape our strategies for living. We discover that there are ideas we are genuinely willing to die for, and more importantly, to live for, and caring for others can play a central role in determining what those are. Moreover, communities that foster such caring are arguably more adaptive to the pressures of natural selection, thereby creating the kind of positive feedback that results in institutionalizing any given strategy for living.

More generally, narratives with analytics are what authorize ethics. The challenge for a secular worldview is that, unlike religion, it provides no master narrative. Instead, akin to an ecosystem of varying species, there is a diverse population of ethical narratives, each focused primarily on securing local validity, each competing to expand its coverage further as circumstances allow. Conflicts are inevitable, and because identities are involved, they can become violent. Such violence, in turn, is a breeding ground for anger, despair, hatred, revenge, and all the other foundational elements of evil.

Evil poses an existential threat to goodness, not least in the temptation to meet evil with evil. To counter this threat, goodness has to play both offense and defense. The latter involves self-sacrifice to curtail the spread of evil as best as one can. The former requires proactively engaging in the world to do good. This is the primary charter of ethics, and the final three chapters focused solely on it.

First we addressed the psychology of ethics, focusing on the ego as a moral agent, acting in collaboration with the rest of the personality as its social representative. In this context, narrative supplies the foundation for moral decision-making, providing strategic direction while allowing for tactical improvisation. Casting the ego as a character acting within a narrative, we counseled it to reject the role of master and adopt the role of servant, both for its own sake and for the good of the personality as a whole. And finally, in an era in which the ego is a frequent target of criticism, we honored the fact that, in making the calls that determine our outcomes, it must act in real time with incomplete knowledge and thereby deserves both our respect and compassion.

Then we turned to the subject of doing good itself. Here metaphysics connects directly with ethics, as doing good sorts out into four domains, each with its own dynamics emerging from a different stairstep in our metaphysical hierarchy. Of these, kindness is the most fundamental, emerging with values as part of mammalian socialization, something we inherit as a birthright prior to any culturalization. Fairness is one of the good tricks that culture provides, once language enables us to conceptualize its principles. Both kindness and fairness operate locally within the confines of family, friends, and community, and as such, are more familiar to us and make up the bulk of our everyday moral decision-making. In contrast, the domains of morality and justice both operate at a global scale, the one emerging from narrative and analytics in the context of specific cultures, the other coming into its own with theory in the context of various nation states. *Global*, however, does not equate to *universal*, and both disciplines struggle with the limits of their authorization—morality when it seeks to impose its narrative onto disbelieving others, justice when it seeks to expand beyond legal justice to social justice. At present, we are still working through these challenges, making each domain a work in progress, in contrast with the surer grounds of kindness and fairness.

Finally, we turned to the ultimate intersection of ethics with metaphysics: mortality. As opposed to demonizing death, we had to admit that we owed it a debt of gratitude, for without it, quite simply, we would not exist. That said, we still had to come to terms with a host of concerns, beginning with finding a narrative that could both inspire and console us, particularly as we and those we love approach end of life. We then looked at how we might adapt our strategy for living to the dynamics of end of life, both part 1, where there are significant new options to exploit, and part 2, where our focus was on closure with dignity.

What is going on? Driven by entropy, the universe is continually emerging, creating increasing complexity, of which life on Earth is a part, and the only part that really matters to us. That complexity creates terraces of activity, each with its own local dynamics, connected to those above and below in ways that are understandable but not deterministic. Our consciousness can embrace the entirety of this ecosystem, from the laws of physics to the heights of theory, and thus we perceive many things going on at the same time, all of which involve us in some way. Given that, what are we supposed to do?

We are born into this world with a compulsion to thrive and reproduce, refined through natural and sexual selection to produce diverse strategies for living. Specifically, we are socially oriented mammals with mutual interests amplified and enabled by our faculty of language. Traditional ethics help us navigate the complex relationships that ensue, seeking to balance personal interests in pleasure and power with altruistic concern for the welfare of others. Historically, ethics have been situated in religious narratives entailing obedience to a divine creator. What we have sought to demonstrate in this book is that they are equally compatible with a strategy for living unfolding in a secular universe. In either case, we are carried forward by the narratives we embrace. They provide the foundation for our strategies for living. We are storytelling animals living out our stories as best we can. That is the common thread that unites us all.

Acknowledgments

When I started writing this book, unlike anything else I have ever written, I did not have a specific audience in mind. Or perhaps better said, I was writing for an audience of one: me. It was just something I had to write, partly to get things straight in my own mind, partly to share whatever bits of wisdom I might have picked up before I pass on. It actually began as a single essay, which then expanded into a series of essays, which then got stitched together by the staircase framework, and continued to develop from there.

I make this point because I want to acknowledge a number of people who engaged with these ideas at various stages, helping both to shape my thinking and sharpen my focus. That work was begun by my son-in-law, Dave Silverstein, who in addition to being a poet, an entrepreneur, a loving husband, and a terrific dad, is also a gifted teacher and editor of writing. His initial comments on each of my chapters were both grounding and encouraging, for which I am deeply thankful. A number of other friends were also enlisted to read various essays as they came into being, each giving me thoughtful critiques of what were emerging ideas. These include Raul Camposano, Elia Van Tuyl, Steve Krause, and my consulting colleague, Todd Hewlin. And then there were three individuals who dug in deeper still to change my thinking and focus in pretty dramatic ways. Both Bob Erickson and Jonathan Dippert helped me get a clearer view of what I was really trying to say about entropy, while Vic Schachter helped clarify my thinking about legal and social justice and the distinction between them, not to mention encouraging me throughout to define my terms.

Under their influence I wrote the first draft of the book you have just read. Normally, that would have put me close to the end of the journey—a few edits, some work on the title, the cover, the blurbs, and *voilà!* Not so in this case. Both my literary agent, Jim Levine, and my publisher, Glenn Yeffeth, weighed in with deep commitment and thoughtfulness. Subsequently, Jim recruited scientist and author Safi Bahcall to review and critique the scientific claims I was making. Between Glenn and Safi, I have never had so many notes on a manuscript in my life, and each one created an occasion to rethink a tenuous claim, bring home a key point with better examples, reframe my relationship to the reader, or simply dig deeper into an idea I thought I had already finished with. I am enormously grateful, not to say lucky, to be teamed with all three of these individuals. And once that work had been done, I found myself in the hands of a terrific copy editor, Karen Wise, who helped me whip the manuscript into final form.

Finally, the person whose support I value most deeply is my wife, Marie. It is not just the patience she has with a writer who is so often in the room but not fully present, nor the care she is taking on behalf of both of us during the recent pandemic, although I am deeply thankful for both. Rather it is the conversations we have, the insights that emerge from them, the ongoing exploration of ideas and values, that mean so much to me, not only as a spouse, but as a writer. In the twenty-first century, it is no longer fashionable for writers to have a muse. That's a pity. I guess I'll just have to be unfashionable.

Bibliography

When you set out to cover as much ground as the book does, you can't possibly research the topic in any disciplined way. Instead, for the past several decades I have been following my heart and reading whatever I thought might interest me. The result is just the sort of hodgepodge you would expect. I share it with you here more in the interest of full disclosure than in providing you a list of further recommended reading. That said, I have

1. grouped the books under the chapter headings where I judge them to have the most relevance, and
2. put in bold the titles that I found most helpful to me.

You may note that far more of the readings relate to Part One than to Part Two. That is because I am trying to weave a whole new story around metaphysics, whereas with ethics, primarily I am seeking to connect traditional ethics to the new metaphysics.

Part One: Metaphysics

Audi, Robert, *The Cambridge Dictionary of Philosophy, Third Edition*

Belsey, Catherine, *Poststructuralism: A Very Short Introduction*

Brockman, John, *This Explains Everything: Deep, Beautiful, and Elegant Theories of How the World Works*

Craig, Edward, *Philosophy: A Very Short Introduction*

Damasio, Antonio, *The Strange Order of Things: Life, Feeling, and the Making of Cultures*

Dennett, Daniel, *From Bacteria to Bach and Back: The Evolution of Minds*

Dewey, John, *Experience and Nature*

Goldstein, Rebecca, *Plato at the Googleplex: Why Philosophy Won't Go Away*

Gottlieb, Anthony, *The Dream of Enlightenment: The Rise of Modern Philosophy*

Gottlieb, Anthony, *The Dream of Reason: A History of Western Philosophy from the Greeks to the Renaissance*

Grayling, A. C., *The Age of Genius: The Seventeenth Century and the Birth of the Modern Mind*

Gura, Philip, *American Transcendentalism: A History*

Harari, Yuval Noah, *Sapiens: A Brief History of Humankind*

Honderich, Ted, *The Oxford Companion to Philosophy*, New Edition

Kaag, John, *American Philosophy: A Love Story*

Kaag, John, *Hiking with Nietzsche: On Becoming Who You Are*

Kaufman, Walter, *Existentialism from Dostoevsky to Sartre*

Kenny, Anthony, *A New History of Western Philosophy*

Menand, Louis, *The Metaphysical Club: A Story of Ideas in America*

Moore, A. W., *The Evolution of Modern Metaphysics: Making Sense of Things*

O'Grady, Jane, *Enlightenment Philosophy in a Nutshell*

Pirsig, Robert, *Zen and the Art of Motorcycle Maintenance*

Randall, Lisa, *Knocking on Heaven's Door: How Physics and Scientific Thinking Illuminate the Universe and the Modern World*

Reed, T. J., *Light in Germany: Scenes from an Unknown Enlightenment*

Royle, Nicholas, *Jacques Derrida*

Scruton, Roger, *Kant: A Very Short Introduction*

Singer, Peter, *Marx: A Very Short Introduction*

Singer, Peter, *Hegel: A Very Short Introduction*

West, Geoffrey, *Scale: The Universal Laws of Life, Growth, and Death in Organisms, Cities, and Companies*

Wilson, Edward O., *The Meaning of Human Existence*

The Infinite Staircase

Bahcall, Safi, *Loonshots*

Carroll, Sean, *The Big Picture: On the Origins of Life, Meaning, and the Universe Itself*

Cziko, Gary, *Without Miracles: Universal Selection Theory and the Second Darwinian Revolution*

Holland, John, *Complexity: A Very Short Introduction*

Johnson, Steven, *Emergence: The Connected Lives of Ants, Cities, Brains, and Software*

Lane, Nick, *Life Ascending: The Ten Great Inventions of Evolution*

Mitchell, Melanie, *Complexity: A Guided Tour*

Morowitz, Harold, *The Emergence of Everything*

Prigogine, Ilya, *The End of Certainty*

Smith, John Maynard, and Eörs Szathmáry, *The Major Transitions in Evolution*

Strogatz, Steven, *Sync: How Order Emerges from Chaos in the Universe, Nature, and Daily Life*

The Metaphysics of Entropy

STAIR 1: PHYSICS: ENTROPY AND THE AFTERMATH OF THE BIG BANG

Ananthaswamy, Anil, *Through Two Doors at Once: The Elegant Experiment That Captures the Enigma of Our Quantum Reality*

Atkins, Peter, *Four Laws That Drive the Universe (And Why Should We Care?)*

Clifton, Timothy, *Gravity: A Very Short Introduction*

Einstein, Albert, *Relativity: The Special and the General Theory*

Ferreira, Pedro G., *The Perfect Theory: A Century of Geniuses and the Battle Over General Relativity*

Fleisch, Daniel, *A Student's Guide to Maxwell's Equations*

Forbes, Nancy, and Basil Mahon, *Faraday, Maxwell, and the Electromagnetic Field: How Two Men Revolutionized Physics*

Gamow, George, *Gravity*

Han, M. Y., *The Probable Universe: An Owner's Guide to Quantum Physics*

Hawking, Stephen, *The Universe in a Nutshell*

Holt, Jim, *When Einstein Walked with Gödel: Excursions to the Edge of Thought*

Krauss, Lawrence, *A Universe from Nothing: Why There Is Something Rather Than Nothing*

Laughlin, Robert B., *A Different Universe: Reinventing Physics from the Bottom Down*

Mahon, Basil, *The Man Who Changed Everything: The Life of James Clerk Maxwell*

Rees, Martin, *Just Six Numbers: The Deep Forces That Shape the Universe*

Rovelli, Carlo, *Reality Is Not What It Seems: The Journey to Quantum Gravity*

Rovelli, Carlo, *Seven Brief Lessons on Physics*

Speyer, Edward, *Six Roads from Newton*

Susskind, Leonard, and George Hrabovsky, *The Theoretical Minimum: What You Need to Know to Start Doing Physics*

Tyson, Neil deGrasse, *Astrophysics for People in a Hurry*

Von Baeyer, Hans Christian, *Maxwell's Demon: Why Warmth Disperses and Time Passes*

STAIR 2: CHEMISTRY: THE EMERGENCE OF EMERGENCE

Aldersey-Williams, Hugh, *Periodic Tales: A Cultural History of the Elements, from Arsenic to Zinc*

McMullen, Chris, *Understand Basic Chemistry Concepts: The Periodic Table, Chemical Bonds, Naming Compounds, Balancing Equations, and More*

Miodownik, Mark, *Liquid Rules: The Delightful and Dangerous Substances That Flow Through Our Lives*

Scerri, Eric R., *The Periodic Table: Its Story and Its Significance*

STAIR 3: BIOLOGY: "...THEN A MIRACLE OCCURS!"

Alberts, Bruce, et al., *Molecular Biology of the Cell, Sixth Edition*

Allen, Terrence, and Graham Cowling, *The Cell: A Very Short Introduction*

Blankenship, Robert E., *Molecular Mechanisms of Photosynthesis*

Bray, Dennis, *Wetware: A Computer in Every Living Cell*

Bryson, Bill, *The Body: A Guide for Occupants*

Davies, Jamie A., *Life Unfolding: How the Human Body Creates Itself*

Dennett, Daniel C., *Darwin's Dangerous Idea: Evolution and the Meanings of Life*

Depew, David J., and Bruce H. Weber, *Darwinism Evolving: Systems Dynamics and the Genealogy of Natural Selection*

Epstein, Randi Hutter, *Aroused: The History of Hormones and How They Control Just About Everything*

Harold, Franklin M., *In Search of Cell History: The Evolution of Life's Building Blocks*

Hoffman, Peter M., *Life's Ratchet: How Molecular Machines Extract Order from Chaos*

Kirschner, Marc W., *The Plausibility of Life: Resolving Darwin's Dilemma*

Kratz, Rene Fester, *Molecular and Cell Biology for Dummies, Second Edition*

Laland, Kevin N., *Darwin's Unfinished Symphony: How Culture Made the Human Mind*

Lane, Nick, *Life Ascending: The Ten Great Inventions of Evolution*

Lane, Nick, *The Vital Question: Energy, Evolution, and the Origins of Complex Life*

McFadden, Johnjoe, and Jim Al-Khalili, *Life on the Edge: The Coming of Age of Quantum Biology*

Mukherjee, Siddhartha, *The Gene: An Intimate History*

Slack, Jonathan M. W., *Essential Developmental Biology, Third Edition*

Sompayrac, Lauren, *How the Immune System Works, Sixth Edition*

Weiner, Jonathan, *The Beak of the Finch: A Story of Evolution in Our Time*

Wolpert, Lewis, *Developmental Biology: A Very Short Introduction*

Yong, Ed, *I Contain Multitudes: The Microbes Within Us and a Grander View of Life*

The Metaphysics of Darwinism

STAIR 4: DESIRE: THE DARWINIAN MEAN

Sapolsky, Robert M., *Behave: The Biology of Humans at Our Best and Worst*

STAIR 5: CONSCIOUSNESS: A DARWINIAN THEORY OF FORMS

Boden, Margaret, *Artificial Intelligence: A Very Short Introduction*
Cobb, Matthew, *The Idea of the Brain*
Eagleman, David, *The Brain: The Story of You*
Fernández-Armesto, Felipe, *Out of Our Minds: What We Think and How We Came to Think It*
Harris, Annaka, *Conscious: A Brief Guide to the Fundamental Mystery of the Mind*
James, William, *The Principles of Psychology*
Kandel, Eric R., *The Disordered Mind: What Unusual Brains Tell Us About Ourselves*
O'Shea, Michael, *The Brain: A Very Short Introduction*
Pinker, Steven, *How the Mind Works*
Ramachandran, V. S., *The Tell-Tale Brain: A Neuroscientist's Quest for What Makes Us Human*
Tomasello, Michael, *A Natural History of Human Thinking*

The Metaphysics of Memes

Blackmore, Susan, *The Meme Machine*
Distin, Kate, *The Selfish Meme: A Critical Reassessment*
Jackendoff, Ray S., *Language, Consciousness, and Culture: Essays on Mental Structure*

STAIR 8: LANGUAGE: THE FABRIC AND FABRICATOR OF MEMES

Bickerton, Derek, *Adam's Tongue: How Humans Made Language, How Language Made Humans*
Everett, Daniel L., *How Language Began: The Story of Humanity's Greatest Invention*
Fish, Stanley, *How to Write a Sentence and How to Read One*
Lakoff, George, and Mark Johnson, *Metaphors We Live By*

Nevalainen, Terttu, and Elizabeth Closs Traugott, *The Oxford Handbook of the History of English*

Pinker, Steven, *The Language Instinct: How the Mind Creates Language*

Portner, Paul H., *What Is Meaning? Fundamentals of Formal Semantics*

Richards, Jennifer, *Rhetoric: The New Critical Idiom*

Rosenfeld, Colleen Ruth, *Indecorous Thinking: Figures of Speech in Early Modern Poetics*

Sacks, Sheldon, *On Metaphor*

Stamper, Kory, *Word by Word: The Secret Life of Dictionaries*

Sullivan, Karen, *Mixed Metaphors: Their Use and Abuse*

Taylor, Charles, *The Language Animal: The Full Shape of the Human Linguistic Capacity*

Turner, Mark, *The Literary Mind: The Origins of Thought and Language*

STAIR 9: NARRATIVE: INVENTING STRATEGIES FOR LIVING

Abbott, H. Porter, *The Cambridge Introduction to Narrative*

Bakewell, Sarah, *At the Existentialist Café: Freedom, Being, and Apricot Cocktails*

Bellah, Robert N., *Religion in Human Evolution*

Booker, Christopher, *The Seven Basic Plots: Why We Tell Stories*

Cascardi, Anthony J., *The Cambridge Introduction to Literature and Philosophy*

Culler, Jonathan, *Literary Theory: A Very Short Introduction*

Frye, Northrop, *Anatomy of Criticism*

Frye, Northrop, *Words with Power: Being a Second Study of "The Bible and Literature"*

Greenblatt, Stephen, *Renaissance Self-Fashioning: From More to Shakespeare*

Greenblatt, Stephen, *The Rise and Fall of Adam and Eve*

Herman, David, Manfred Jahn, and Marie-Laure Ryan, *Routledge Encyclopedia of Narrative Theory*

Herman, David, *The Emergence of Mind: Representations of Consciousness in Narrative Discourse in English*

James, William, *The Varieties of Religious Experience*

McKee, Robert, *Story: Substance, Structure, Style, and the Principles of Screenwriting*

Spacks, Patricia Meyer, *On Rereading*

White, Hayden, *Metahistory: The Historical Imagination in Nineteenth-Century Europe*

STAIR 10: ANALYTICS: TESTING AND REFINING STRATEGIES FOR LIVING

Alcock, Lara, *Mathematics Rebooted*

Alpaydin, Ethem, *Machine Learning: The New AI*

Baggini, Julian, and Peter S. Fosl, *A Philosopher's Toolkit: A Compendium of Philosophical Concepts and Methods*

Berlinski, David, *A Tour of the Calculus*

Chang, Ha-Joon, *Economics: The User's Guide*

Domingos, Pedro, *The Master Algorithm: How the Quest for the Ultimate Learning Machine Will Remake Our World*

Felski, Rita, *The Limits of Critique*

Guillen, Michael, *Five Equations That Changed the World: The Power and Poetry of Mathematics*

Haigh, John, *Probability: A Very Short Introduction*

Kahneman, Daniel, *Thinking, Fast and Slow*

Livio, Mario, *The Golden Ratio*

Maor, Eli, *e: The Story of a Number*

Morris, Dan, *Bayes' Theorem: A Visual Introduction for Beginners*

Orlin, Ben, *Math with Bad Drawings: Illuminating the Ideas That Shape Our Reality*

Seife, Charles, *Zero: The Biography of a Dangerous Idea*

Sowell, Thomas, *Basic Economics: A Common Sense Guide to the Economy*

Spiegelhalter, David, *The Art of Statistics: How to Learn from Data*

Toulmin, Stephen, *The Uses of Argument*

Valiant, Leslie, *Probably Approximately Correct: Nature's Algorithms for Learning and Prospering in a Complex World*

van Eemeren, Frans H., and Peter Houtlosser, *Dialectic and Rhetoric: The Warp and Woof of Argumentation Analysis*

Wetherbee, Jim, *Controlling Risk: Thirty Techniques for Operating Excellence*

STAIR 11: THEORY: ONE MEME TO RULE THEM ALL

Buchanan, Ian, *Oxford Dictionary of Critical Theory*
Caldarelli, Guido, *Networks: A Very Short Introduction*
James, William, *Pragmatism*
Pepper, Stephen, World Hypotheses

Being: A Bridge to Ethics

Yogi, Maharishi Mahesh, *Science of Being and the Art of Living*

Part Two: Ethics for the 21st Century

Becker, Charlotte B., *A History of Western Ethics, Second Edition*
Haidt, Jonathan, *The Righteous Mind: Why Good People Are Divided by Politics and Religion*
Ignatieff, Michael, *The Ordinary Virtues: Moral Order in a Divided World*
MacIntyre, Alasdair, *After Virtue: A Study in Moral Theory, Third Edition*
Tomasello, Michael, *A Natural History of Human Morality*
Wootton, David, *Power, Pleasure, and Profit: Insatiable Appetites from Machiavelli to Madison*

Understanding Goodness

Haidt, Jonathan, *The Righteous Mind: Why Good People Are Divided by Politics and Religion*

Honoring the Ego

Stevens, Anthony, *Jung: A Very Short Introduction*

Doing Good

Beeman, Richard, *The Penguin Guide to the United States Constitution*

Haidt, Jonathan, *The Righteous Mind: Why Good People Are Divided by Politics and Religion*

Rawls, John, *A Theory of Justice: Original Edition*

Being Mortal

Gawande, Atul, *Being Mortal: Medicine and What Matters in the End*

Kalanithi, Paul, *When Breath Becomes Air*

About the Author

Photo by Claudia Goetzelmann

GEOFFREY MOORE is an author, speaker, and high-tech business adviser best known for his seminal book, *Crossing the Chasm*, first published in 1990. The book is still in print, having sold over a million copies and been translated into over a dozen languages, and is still required reading in most business schools. Since then, he has published six other bestselling business books and advised leading high-tech enterprises including Salesforce, Microsoft, Intel, Adobe, Autodesk, Box, Airbnb, and Splunk.

Moore has a bachelor's degree in American literature from Stanford University and a PhD in English literature from the University of Washington, where his dissertation was on strategies for living in Spenser's *The Faerie Queene*. Subsequently, he taught literature, conceptual models, and writing for four years at Olivet College before returning to the Bay Area to take up a career in business.